understanding the pattern of life

understanding the pattern of life

ORIGINS AND ORGANIZATION OF THE SPECIES

Todd Charles Wood and Megan J. Murray

Kurt P. Wise, General Editor

Illustrated by Tracy J. Daugherty

Nashville, Tennessee

0-8054-2714-7

Published by Broadman & Holman Publishers,
Nashville, Tennessee

Dewey Decimal Classification: 231.765
Subject Heading: LIFE–ORIGIN \ CREATION

1 2 3 4 5 6 7 8 9 10 07 06 05 04 03

Contents

Preface

I pray that what you are about to read will be unlike any creationist book that you have ever read. It has never been my intention to write yet another book about the truth of Scripture or perceived inadequacies of evolution. Instead, this book is an expression of my attempt, however feeble, to allow the truth of God to transform my view of biology. Rather than trying to prove the truth of Scripture, I assume it as a starting point. From there, I build what I believe to be a reasonable model of biology that fits both the facts of Scripture and the data of creation. You will find that evolutionary theories are mentioned only rarely, and when I do discuss them, I do so to highlight the differences between my ideas and the conventional view.

To the skeptics who will no doubt read this book in order to destroy it, allow me to save you much trouble. The book is unabashedly "religious" in tone. I wholeheartedly reject the belief that religion and science occupy separate domains. I believe that the Bible is the infallible Word of God to man, true in every detail of the original autographs. Whether it speaks on the fate of my eternal soul or on the fate of Noah and the animals during the Great Flood, it is equally true. Throughout this book, I will consistently check my interpretation of science against the Bible as my standard of truth. I will constantly quote from its passages to illustrate the coherence of my theories and interpretations with the Holy Writ.

And best of all, I make no apology for this perspective, for the Word of God is quick and powerful and sharper than any two-edged sword. The Word of God is the reason I am a creationist in the first place; why should I ignore it when I do my science? I have no time to defend my view, and I have little need to do so. If the Word of God is truth, as I believe it is, how could I, a mere fallen mortal, ever hope to defend it? That indeed would be a waste of time, and the task before me is too great for such distractions.

To my creationist readers, I issue a gentle warning. This book is not an apologetic of any kind. If you are looking for evidence for your faith, I offer none other than the Bible itself. God's Word is all the reason to believe that you or I could ever need. This book is a technical presentation of a foundational field of creation biology. I deal with hard questions in these pages, and more often than not, I do not have the answers. Sometimes I will even reject common creationist opinion and practice, primarily because they simply don't work. The best I can hope is to clarify the questions for future generations. Rest assured, however, that I believe that the answers are forthcoming if we faithfully keep God as our central focus in science. After all, creation is an expression of His very nature, so how could we do science without Him?

It is very important for all readers to realize that the synthesis of ideas presented in this text, like all scientific theories, is a work in progress. I have attempted to draw on the tradition of creation biology as much as I can while at the same time consulting experts in numerous fields to make sure that my ideas are sound. Since I am the primary author of most of these ideas (my coauthor is a teacher and writer but not a scientist), any errors in fact or logic that you may discover are mine alone. Do bring them to my attention so that I may correct them in future revisions of the book.

Finally, I want to address my primary audience, the Christian student of biology. To those of you still confused about the whole creation/evolution issue, I hope that this book will show you that real science is possible without rejecting a straightforward reading of Genesis. My prayer is that you will find something of value here in your own search for answers. To those of you who share my view of Scripture, I present the "state of the art" of creation biology. This is where we are as of A.D. 2003, and I have tried to outline where we need to go next. I pray that this book will serve both as a retrospective and a guide to creation biology research. Above all, I hope that this book will be an inspiration to a new generation of creation biologists who will inherit some of my passion for God's living creation.

Todd Charles Wood

SECTION I

Foundations

CHAPTER 1

Foundations of Baraminology

1.1. Introduction

Everywhere we look we encounter a dizzying array of living things. From the dogs and cats we keep as pets, to the trees and grass we treasure in our parks, to the mildew and fungus we try to keep from our homes, God's creatures are everywhere. We could easily spend several lifetimes examining the creatures on the earth, but we would still not even come close to discovering the entirety of biological diversity. With such an amazing variety of organisms available to research, biologists have devised methods to study, name, and classify creatures. **Baraminology** is one such method.

Baraminology is a creationist method of **biosystematics**. As distinguished from **taxonomy**, biosystematists (or systematists for short) deal with the discovery and naming of real or natural groups of biodiversity. Taxonomists try to classify diversity into larger groups. Unfortunately, taxonomy and systematics are often confused colloquially, but they are distinct activities. Conventional systematists group organisms into **species**, and conventional taxonomists arrange species into groups (families or phyla, for example), which by common admission are somewhat arbitrarily defined. Because creationists advocate real groups that contain multiple species, baraminology resembles both biosystematics and taxonomy. In this introductory chapter, we will present a generic history of classification in order to lay the conceptual and historical foundations for understanding baraminology. We will return to the shades of meaning inherent in *baraminology, systematics,* and *taxonomy* in chapter 2.

1.2. What's in a Name?

Biosystematics is all about words. These words often sound strange to our modern ears: *Thermoplasma acidophilum, Ureaplasma urealyticum, Giardia lamblia, Latimeria chalumnae, Gossypium hirsutum.* Systematists take very seriously the naming of groups of creatures, by tradition using Latin or latinized words. Committees of systematists design and maintain the many formal rules of **nomenclature**, the way we name organisms. These rules are published by numerous scientific organizations dedicated to biological nomenclature.

For example, the 1999 *International Code for Botanical Nomenclature* had three editors and 475 pages all devoted to the proper naming of plants.[1] With all this effort focused on nomenclature, many people, biologists included, wonder if systematics might be a bit too obsessive. Categorization and naming certainly make biology more convenient, but should scientists really bother arguing over the category and name of every insignificant creature on the planet? Even the categories themselves can be contentious. Witness the quest to create a "**species concept**": There have been dozens upon dozens of attempts by scientists to define exactly what a species is. Is good biological naming really worth all this effort? Since this book is ultimately all about biosystematics, we shall begin by addressing this question of the importance of naming.

If we view systematics and nomenclature as nothing more than dispensing arbitrary designations, like license plates or social security numbers, quibbles over names do seem silly, but nomenclature goes far beyond mere designation. To name something well is to know the thing and to see it in a different way. The names we give and the descriptions we make change us by opening our eyes to the world around us. Ultimately, the names we give both reflect and transform our view of ourselves and our view of God. Because of this power intrinsic to nomenclature, we should take biosystematics very seriously. With a good, godly biosystematics method, we have the potential to completely revolutionize our view of all of biology.

The Bible says a great deal about the value and power of names and the words we speak. The Book of Proverbs depicts the power of the tongue very explicitly: "Death and life are in the power of the tongue" (Prov. 18:21). The power of the tongue is further expounded by James. He writes, "Behold, we put bits in the horses' mouths, that they may obey us; and we turn about their whole body. Behold also the ships, which though they be so great, and are driven of fierce winds, yet are they turned about with a very small helm, whithersoever the governor listeth. Even so the tongue is a little member, and boasteth great things" (James 3:3–5). We tend to relegate the meanings of passages like these to the realm of personal relationships. As James writes, "Out of the same mouth proceedeth blessing and cursing. My brethren, these things ought not so to be" (James 3:10). James clearly speaks of the moral aspect of using words, but the underlying power of words impacts all areas of life.

Unfortunately, the power of words is most clearly seen in their power to destroy. The words of Karl Marx and Adolf Hitler left indelible scars on the face of the twentieth century. Imagine what the world would have been like if *Mein Kampf* or *Das Kapital* had never been written. Many citizens of the United States still shudder at the mention of Senator Joseph McCarthy. *Communist* is a mere word, but when uttered from the mouth of Senator McCarthy, that word irrevocably damaged the lives of many people. Words of anger and hate uttered by Osama bin Laden left a crater in Manhattan and in the hearts of thousands who lost loved ones at the World Trade Center.

If bad words accomplish such chilling and unspeakable evil, just think what good words can do. Proverbs states, "A wholesome tongue is a tree of life" (Prov. 15:4). The apostle John recognized the power of words in the life of Christ. Sixty years after Jesus returned to heaven, John wrote his Gospel, the latest of all four evangelists. John undoubtedly spent much time in those sixty years pondering the life of Jesus, and he chose to open his Gospel by describing Christ as the Word (John 1:1–3, 14). Of all the pictures John could have conveyed, he chose to represent the meaning and power of Jesus Christ as a Word. Considering the power of words, what could be more appropriate?

So we should not regard the task of naming as something trivial or unimportant to the larger field of biology. We should strive to honor the Savior and Creator with our words in all areas of life. To the Christian systematist, naming organisms is not merely a scientific exercise. These words can transform our view of God and **creation**. If done wisely and with proper respect to God's creation, our naming of organisms can become a "tree of life," a source of praise to God and encouragement to our fellow believers. With the psalmist, we ought to pray, "Let the words of my mouth, and the meditation of my heart, be acceptable in thy sight, O LORD, my strength, and my redeemer" (Ps. 19:14).

1.3. Biosystematics

As is the case for so many things, biosystematics finds its origin in the creation account of Genesis.

> And the LORD God said, It is not good that the man should be alone; I will make him an help meet for him. And out of the ground the LORD God formed every beast of the field, and every fowl of the air; and brought them unto Adam to see what he would call them: and whatsoever Adam called every living creature, that was the name thereof. And Adam gave names to all cattle, and to the fowl of the air, and to every beast of the field; but for Adam there was not found an help meet for him. And the LORD God caused a deep sleep to fall upon Adam, and he slept: and he took one of his ribs, and closed up the flesh instead thereof; and the rib, which the LORD God had taken from man, made he a woman, and brought her unto the man. And Adam said, This is now bone of my bones, and flesh of my flesh: she shall be called Woman, because she was taken out of Man. (Gen. 2:18–23)

As scientists, we might find this passage peculiar at the least, almost whimsically mythical, but as Christians, we must remember that "All scripture is given by inspiration of God, and is profitable" (2 Tim. 3:16). What profit does this passage have for the systematist? By examining these verses, we can find four important biosystematics principles. Biosystematics is a valuable, natural, intuitive human activity.

1.3.1. Biosystematics Is Valuable

Scientists expend a lot of words trying to justify their research, mostly while writing grant proposals. A justification of systematics research typically invokes the importance of biodiversity to a healthy ecosystem or the potential for therapeutic discoveries in hitherto unknown creatures. As Christians, we can look beyond these pragmatic concerns to a richer justification rooted in God Himself, as revealed in Genesis 2:18–23.

After placing the newly created man in the Garden of Eden "to keep it" (Gen. 2:15), God called Adam to examine and name the things He had made. By itself, God's command lends meaning and purpose to our activity. If God deems it valuable enough to call Adam to do it as one of his first tasks, so should we. Beyond simple obedience, we must also recognize the importance of examining God's creation as His original revelation of His nature to us. By systematically studying the work and Word of God, we come to know Him better. We can recognize God's "glory" and "handiwork" (Ps. 19:1) and His "eternal power and Godhead" (Rom. 1:20). This then is the ultimate goal of Christian biosystematics: not merely to revel in creation but to worship and honor the Creator.

As we come to know God better through biosystematics, everything we think we know about biology will become transformed. Perceived difficulties will disappear, and new challenges will come to light. As our view of biology changes, the practice of creation biologists will change too. New and uniquely creationist avenues of research will open up, ultimately yielding a thriving science of creation biology. The last section of this book will illustrate how biosystematics changes everything.

1.3.2. Biosystematics Comes Naturally

When God brought the birds and beasts to Adam to see what he would call them, He did not teach Adam what to do. Apparently, Adam knew exactly what to do, and we still see the human proclivity for classification and naming alive and well today. In everything we do, for good or for ill, we classify and organize the things and people around us. No one needs to teach us how to do this. It just comes naturally.

We gain further insight into the human ability to classify and name from the only recorded names bestowed by Adam. In Genesis 2:23, Adam gives his wife the category name "woman." After the Fall, he names his wife with the personal name "Eve" (Gen. 3:20). We also have the same tendency to give and use general category names and specific personal names. From the cars we drive–Ford (category) Mustang (specific)–to the clothes we wear–cardigan (specific) sweater (category)–dual names permeate our culture. Remember, too, that category names and personal names need not be specific. A modern woman may find herself in many categories (teacher, author, dog owner) and may carry more than one personal name (first name, middle name, nickname). Similarly, personal names can apply to more than one person. Many people carry the names Bill, John, Susan, or Mary. Individuals can carry many names, and names can apply to many people.

Even though the classification of biosystematics operates on formal rules, we can learn how to formulate these rules from the natural and informal classifications that we use all the time. By observing our everyday classifications, we infer an important theme: Classification changes according to our need. Suppose a corporate executive needs to meet with his sales department. He would address his memo to "Sales Personnel" rather than listing their names. Suppose the same executive needs to meet with the New York office. That could include everyone from the janitor to the head of the office, and it would overlap slightly with the sales personnel. Thus, we keep our casual classifications fluid so that they will be useful to us.

Figure 1.1. *People often give objects dual names. Though all of these objects are bowls, they differ according to their purpose. Reflecting those differences in use, each bowl has a different specific name: punch bowl, mixing bowl, soup bowl, candy bowl.*

This same fluidity of classification is found in the Bible. When Paul wrote to the Galatians to refute challenges to his authority, he wrote as "Paul, an apostle, (not of men, neither by man, but by Jesus Christ, and God the Father, who raised him from the dead)" (Gal. 1:1). When he wrote to encourage the Philippians, he addressed them as a servant of Jesus Christ (Phil. 1:1). The need changed from church authority figure correcting doctrinal error ("apostle") to fellow believer excited about life with the risen Christ ("servant"). Paul simply referred to a different classification of himself as the situation warranted.

In modern biosystematics, scientists emphasize a unique, true classification of organisms. In this system, species are grouped into unique genera, which are grouped into unique families, and so on. Because this rigid system allows no overlap, any given species can be a member of only one genus and only one family. Considering how human beings usually construct and use classifications, we see that the conventional biological system actually runs counter to our experience with the useful, fluid classifications of our daily experience. Thus, the creationist systematist should wisely refrain from insisting on a single, "natural" classification.[2] We will return to this theme in the following sections.

1.3.3. Biosystematics Is Intuitive

Among the many skeptical criticisms of the Genesis creation account, Adam's naming of the animals looms large. Critics claim that there is not enough time for Adam to name every species of animal in a single twenty-four-hour day. If we look at the same verse (Gen. 2:19) from the perspective of scriptural inerrancy, we come to a very different understanding. Rather than being impossible, Adam's naming of the birds and beasts reveals the important truth that our classification is intuitive.

First of all, the verse does not claim that Adam named all members of the modern kingdom Animalia. It says that he named "every beast of the field" and "fowl of the air" (Gen. 2:20 includes "cattle"). Creatures of the ocean are not included, nor are "creeping things" (cf. Gen. 1:24–25). Second, the verse records no mention of the precise groups that he named.

Individual organisms, species, genera, families, even phyla are legitimate possibilities in the face of this scriptural silence. If Adam named only families or orders of the birds and beasts, he could easily accomplish his task in an afternoon. We must be careful not to impose our modern, scientific mind-set on the meaning of Scripture. More often than not, critiques of inerrancy arise from this unfortunate habit.

Whatever the groups were that Adam named, he seemed to do his job without any instruction, as we noted above. Adam's classification in Genesis 2:19 sharply contrasts with modern biosystematics. Systematists today employ careful studies of the characteristics of organisms coupled with complex mathematical algorithms to distinguish groups of organisms. In contrast, Adam seemed to know the groups intuitively. The Bible records no careful examinations or complicated calculations. Adam intuitively recognized the groups and named them.

Adam's intuitive classification should not in any way invalidate or belittle modern systematics methods. Calculating statistical significance for a particular grouping of organisms lends credence to our systematic hypotheses. Instead, Adam's intuition stands as a warning against **reductionism**. When Adam named the beasts and birds, he did not examine their genetic similarity, nor did he search for a single, definitive characteristic for a group. Adam merely looked at the whole animal and knew precisely what to call it. In developing biosystematic methods, the creationist systematist should strive for holistic techniques that represent organisms as complete creatures. Reducing creatures to a single trait or gene will distort our classifications.

1.3.4. Biosystematics Is a Human Activity

Normally, we take for granted the human participation in biosystematics, but when Adam named the birds and beasts, the possibility of a divinely-revealed classification should be considered. If there exists a "correct" classification scheme and if knowledge of that scheme is important, God should have simply told Adam the names of the creatures. Why did Adam do the naming? One might reply that God or Adam or both needed some kind of a test of Adam's new intellect, but it is difficult to imagine why God or Adam would need to test the "very good" creation. Another possibility is that God revealed the true classification to Adam in Adam's mind. The intuitive classification of Adam came not from his own perception but from divinely created memories of God's true classification.

Either of these answers could contain elements of truth, but we ought to consider a third, more likely possibility: To name is to attempt to understand, and to understand is to better know God. Thus, God's command to name the birds and beasts is actually an invitation to the greatest task of all-knowing God. As we come to know God through His creation, we must recognize that the complexity of the divine plan for biological creation forever eludes complete description by human classification.[3] Human grouping of organisms cannot capture or portray the entirety of God's plan of creation. When God called Adam to name the animals, God knew that Adam's names could cover only part of His plan. We must consider this possibility because of the complexity of God's own nature.

For example, Christians confess that God exists as three persons in one. The Spirit serves the Son who serves the Father, yet all three are equally God. Can anyone understand this mystery? Likewise, can we truly describe it in words? Even when we make up words to describe God, they contain inherent contradictions. *Triune* means "three" and "one" at the same time. The mystery of God is that He is neither three gods nor one person, yet He is both three and one.

Because we know that God's nature is revealed to us in His creation (Ps. 19:1; Rom. 1:20), we should expect that His plan of creation will exhibit a similarly mystifying complexity that just might defy simple description. Thus, the fluidity of human classification might not arise from mere convenience or utility. Our multiple, overlapping, contradictory classifications might simply reflect the complexity of God's plan of creation. If this is the case, we must remember not to insist that a particular human-created classification of organisms is the "correct" one, for other systems may just as accurately describe God's plan of creation. God's call for Adam to name the animals was an invitation to a lifetime of studying God's plan.

The possibility of an inconceivable plan of creation raises an important question: If God's plan is indescribably complex, should we waste our time trying to describe it? To answer this question, we can look at the doctrine of the **Trinity** again. If anything exists beyond human comprehension, surely it is God's triune nature; nevertheless, church fathers at the Council of Nicea codified the truth of Scriptures into the doctrine of the Trinity, a description of God's nature. They did this in part to correct heresy in the church. Another important aspect of the Trinity doctrine is that it is not wrong. To say that there is only one God is not wrong. It is part of the whole truth, and as long as we acknowledge its incompleteness, it remains true. When we insist that "one God" encompasses the whole truth excluding "three persons," we stray from truth to error. Finally, we may ask a counter question: If God's nature defies human scrutability, why does He keep revealing it to us?

Relating these points to biosystematics, we find three important principles. First and perhaps least important, a good statement of the principles of classification is required to counter the error of purely phylogenetic classification. Second, any classification system reflects *part* of the truth but not the *whole* truth. Classifying organisms as birds and mammals is no less "real" or "true" than classifying them as sea creatures, land creatures, and flying creatures. Each system reflects part of the truth. The last, best reason to engage in biosystematics is God's invitation to do so. Even though God knew that His plan for creation exceeded any simple description by Adam, He still brought the birds and beasts to Adam "to see what he would call them" (Gen. 2:19). We continue to honor and delight God by investigating and naming the creatures that He has made.

1.3.5. The Biblical Basis of Biosystematics

From our study of Genesis 2:18–23, we have seen that biosystematics forms a crucial activity by which humans come to know the Creator better. The picture of biosystematics depicted in Adam's naming of the animals differs markedly from modern conventional systematics in several respects. Most importantly, the value of systematics comes from God's invitation to know Him better by studying His creation. Beyond the mundane questions of methodology or utility, biosystematics transforms biology by transforming and renewing the minds of biologists. As we begin this task of transformation, we should remember the lessons of Adam's classification. While modern systematics has turned to reductionism to force organisms onto a unique evolutionary tree, Adam viewed animals holistically with no *a priori* commitment to any particular arrangement.

Finally, since the nature of God is both infinite and inconceivably complex, we should expect that His creation will reflect a similar complexity. Just like studying God's Word, studying God's creation can occupy the lifetime of the committed Christian biologist.

1.4. A Brief History of Biosystematics

The development of modern biosystematics[4] began in ancient Greece with Aristotle. For good or bad, much of his philosophy still resonates in modern systematics. Vestiges of Aristotle's **essentialism** and his "**chain of being**" can still be found in modern creation and **evolution** theory. Beginning with Aristotle, the history of biosystematics in the English-speaking world has swung between two conceptual extremes. Aristotelian **fixity** of species lies at one extreme with Darwin's **universal ancestry** at the other. As we shall see, the truth resides somewhere in the middle, exhibiting characteristics of both extremes. Here again, we find that the truth of God's creation resists monolithic classification.

For brevity's sake, we will limit our discussion of historical systematics to the major influences on modern creationist biosystematics, necessarily ignoring important developments in biology. Beginning with Aristotle, we will trace the history of essentialist fixity through the Reformation up to and including Linnaeus and Cuvier. Very close on the heels of Linnaeus came the **transformists**, such as Geoffroy St. Hilaire and Lamarck, who believed that one species could transform into a different species. Transformism won the minds of intellectuals everywhere by the end of the nineteenth century, thanks to Darwin's *Origin of Species*. We will conclude our discussion with twentieth-century figures important to creationist biosystematics, particularly Frank Lewis Marsh.

1.4.1. Aristotle (384–322 B.C.)

A student of the philosopher Plato, Aristotle[5] studied and wrote in what we now call the field of zoology. Although students of biology undoubtedly existed before him (Solomon was noted for his knowledge of the subject; see 1 Kings 4:32–33), scholars consider Aristotle the first biologist because he produced a written record of his system. Aristotle's views on biology reflect the dominant Greek worldview of his time and should not be evaluated from a strictly modern perspective. Despite the distance of history and culture, his philosophy contains elements that strongly influenced the traditional position of creationists and contributed to the quick acceptance of evolution. Aristotle's belief in the

Figure 1.2. *Aristotle (384–322 B.C.)*

immaterial, immutable essence of organisms still appeals to creationists and repulses evolutionists. At the same time, creationists recognize that Aristotle's "chain of being" has an instinctive appeal to the evolutionary mind-set. Even though many modern creationists and evolutionists would not adhere to either of these ideas, they still affect the way many people think about biology. The complexity of Aristotelian biology will help us to understand why Aristotle has become so frequently vilified.

According to Aristotelian biology, our ability to recognize groups of organisms comes from our perception of certain essential and constant features present in all members of the group. This idea is called essentialism and forms the foundation of Aristotelian epistemology. According to Aristotle, we could not know anything without essences. Since the essences form the basis of our recognition of things around us (not just animals), we would be unable to recognize or know anything about the world if essences could change. What we knew about an object would be invalidated the moment the essence of that object changed. Aristotle wrote,[6] "The object of scientific knowledge is of necessity. Therefore it is eternal: for things that are of necessity in the unqualified sense are all eternal." As we shall see, essentialism became an important ingredient in early species concepts through the work of the medieval **scholastics**.

In addition to essentialism, Aristotle's "chain of being," or ***Scala Naturae,*** continues to influence biology today. To derive this chain, Aristotle began with the four classical Greek elements–earth, water, wind, and fire–and on the basic characteristics of these elements–hot, cold, moist, and dry. Aristotle ranked these characters in order of their importance, with hot outranking cold and moist outranking dry. He viewed blood, being hot and moist, as a very important feature of organisms. Animals with blood were more important than those without. Following this reasoning, Aristotle produced a linear rank of organisms. Among organisms, Aristotle placed worms and flies at the bottom of the chain and human males at the top. Below flies and worms were various inanimate objects and above human males were the celestial beings. This apparent progression of organisms from simple to complex will appear again in transformist ideas.

1.4.2. Scholasticism

The rediscovery of Aristotle during the Middle Ages precipitated a fusion of his philosophy with the Christian faith of the day, resulting in a philosophy/theology hybrid called scholasticism.[7] Chief among the scholastics was Thomas Aquinas (1225–74), a Dominican monk and professor of theology. In his *Summa Theologica,* Aquinas explicitly and extensively employed Aristotelian logic and epistemology in his justification of Christian theology, even though he disagreed with some points of Aristotle's metaphysics. With his famous "Five Ways," he set out five proofs of the existence of God using reason alone. His high opinion of Aristotle's philosophy stemmed from his belief that the Fall of man into sin affected only the will of man but not the intellect. Consequently, Aquinas confidently affirmed that theology could be established through reason, with minimal information from the Scripture.

The Reformers brought a brief lull to the scholastic reverence of Aristotle. For example, Martin Luther disdained the broad authority granted to Aristotle in Christian theology. Luther found it especially outrageous to argue for Christian theology from human reason.

He strove to establish a Christian faith based on the divine revelation of Scripture alone (*sola scriptura*) and cited Genesis 6:5 as evidence of the depravity of human intellect as well as human will. Despite these sometimes vehement objections to scholasticism, the Protestants soon returned to Aristotelian philosophy to express and defend their faith. For example, Melanchthon condemned Aristotle during his early career, but after Luther's death, he actually praised Aristotle as a wise man who was naturally aware of God. Similar commendations and utilization of Aristotle can be found extensively in Protestant writings less than a century after Luther.

The ramifications of the debates surrounding Aristotle and scholasticism during the Reformation were not limited to theology. As we will see in the following section, the dominance of scholasticism and the great respect afforded to Aristotelian philosophy strongly influenced many other developing fields, including biology and systematics.

1.4.3. Linnaeus (1707–78)

Known today as the father of taxonomy, Carl von Linné (Latin: Carolus Linnaeus)[8] was born in Stenbrohult in southern Sweden to the family of the local Lutheran pastor. Linnaeus traveled in 1735 to the Netherlands, where he earned his medical degree from the University of Harderwijk. Upon returning to Sweden, he was appointed to a professorship in botany at Uppsala University. During his life, he produced several seminal works in botany and taxonomy, including *Philosophia Botanica* and the twelve editions of *Systema Naturae*.

Figure 1.3. *Carolus Linnaeus (1707–78).*

By the time of Linnaeus, Aristotle's essentialism held sway over much of science. In the area of biology, most believed that species were composed of individual organisms that possessed a common set of essential (immutable) characteristics. Unlike Aristotle, most biologists attributed the origin of species to the direct act of creation by God. Early in his life, Linnaeus wholeheartedly endorsed this position with his famous statement, "We count as many species as different forms were created in the beginning."

Linnaeus also believed in a logically ordered creation, estimating the existence of only ten thousand species of plants and the same number of animals. To Linnaeus, the purpose of classification was the organization of biological information; thus, he created a higher classification based on simple logical divisions. Linnaeus's innovation came from his choice to restrict his classification to only four levels: class, order, genus, and species. Prior to Linnaeus, biologists used dichotomies to classify organisms (e.g., animals having blood vs. bloodless animals). Dichotomous classification schemes became cumbersome and ultimately unusable as more and more species were added. With Linnaeus's system, classification became practical again.

Linnaeus spent much of his life examining animal and plant specimens from around the world and listing their genus and species names in his *Systema Naturae*. This habit of dual names became a biosystematic standard that we still practice today (**binomial nomenclature**). During his examination of specimens, Linnaeus encountered several organisms that represented anomalies to his essentialist view of immutable species. Interspecific hybrids posed a particularly difficult problem for him. Soon after studying hybrids, Linnaeus began to suspect that species were not immutable. He tried to accommodate hybrids by proposing that genera were the fixed units of God's creation. By **hybridization** between genera, individual species arose.

The problem of hybridization did not go unnoticed by other biologists. In 1787, the prominent English naturalist John Hunter retained Linnaeus's earlier essentialist species concept. To explain canid hybrids, he merely included wolves, dogs, and jackals in the same species.[9] Kölreuter actually performed hybridization experiments and showed that his hybrids did not become new species. From this work, he argued that Linnaeus was wrong to abandon the essentialist species concept. He argued that since hybrids between species could often produce no offspring, their sterility actually reinforced the identity of the parent species.

We must pause here in our history to emphasize a crucial point. Many modern Christians and skeptics alike believe that the Bible teaches species immutability (also called species stasis). As we have seen in this brief overview, the primary impetus for the concept of species stasis comes from essentialism. Even though individual creationists have rejected species stasis for sixty years now, criticism and confusion continues. Thanks to scholastic theology, the mingling of Christianity and Aristotelianism still biases our reading of Scripture. In order to create a truly biblical biosystematics, we must critically and skeptically re-evaluate essentialism in the light of biblical truth. We ought to practice some of the creativity displayed by Linnaeus and Hunter as they tried to modify their scientific theories without abandoning the authority of Scripture.

1.4.4. Time of Transition

Ironically, although we have referred to "biology" and "biologists" throughout the preceding sections, the actual word *biology* was not coined until 1800.[10] With the new moniker came theoretical upheaval during the close of the eighteenth century and the first half of the nineteenth century. Interspecific hybrids were just the beginning, as numerous observations and ideas contributed to the reshaping of biology. Some researchers during this period proposed that species could change into other species in violation of the essentialist dogma. French and German biologists challenged anatomical theories that relied exclusively on function to explain organismal features. The explanatory power of Newton's mechanistic physics became a goal for theoretical biologists. To illustrate the diversity of opinion, we will briefly survey the work of three French biologists: Lamarck, Cuvier, and Geoffroy.

Jean Baptiste Lamarck (1744–1829) is chiefly remembered today as a transformist and forerunner to Darwin. Even though he held a professorship at the Musée National d'Histoire Naturelle (MNHN) in Paris, his transformist views were poorly received during his lifetime, and he died a pauper. In a radical departure from Aristotelian essentialism, Lamarck believed that species constantly adapt to the changing environments that they occupy. At the same time, Lamarck also seems to have endorsed a transformist version of the "chain of being" by his linear view of species transformation. In the Lamarckian view, species progress from simple to more "perfect" forms as they change with their environments.

One of Lamarck's chief critics, Georges Cuvier (1769–1832) also held a professorship of anatomy at the MNHN. Cuvier departed very slightly from Aristotle in his belief that species could become extinct; Aristotle would have rejected extinction because essences are eternal. On most other substantive issues, Cuvier followed Aristotle closely. Cuvier envisioned each extant species as the descendants of a single pair that was directly created by God, optimally designed for the environments they occupy. Cuvier rejected transformism partly because of his high regard for functional design and partly because of his Aristotelian epistemology. He believed that allowing the transformation of species was equivalent to destroying the very basis of scientific knowledge. Despite challenges posed by transformists, Cuvier remained forever unshaken in his belief in species stasis.

A third prominent professor at the MNHN, Éttiene Geoffroy St.-Hilaire (1772–1844), believed in the primacy of form over function in biology. Rather than focusing on the functional adaptation of organisms to their environment, he believed that animals shared a common body plan, a view known today as **typology** or **structuralism**. He devoted his research to understanding what he perceived to be morphological laws that governed animal form. He also believed in transformism, though he never developed a system as sophisticated as Lamarck's. His advocacy of biological similarities (later called **homologues**) between vertebrates and invertebrates drew sharp criticism from Cuvier and precipitated their famous debate in 1830. Though Cuvier was considered the winner of the debate, he was ultimately unable to stop the spread of transformism.[11]

1.4.5. Charles Darwin (1809–82)

Growing up in this period of theoretical turmoil, Darwin ultimately won the day with his theory of evolution, in many respects the polar opposite of Aristotelian biology. Many modern evolutionists regard Darwin's theory as a victory for the functionalists. Darwin believed that natural selection shaped organic form by favoring organisms that were better adapted to their environment. Thus, the major determinant of animal morphology was how well that morphology enhanced survival of the offspring. In this way, Darwin explained the "perfection" of form to function that so captivated Cuvier. As we shall see, this view of Darwin as functionalist necessarily ignores an important aspect of his theory.

Figure 1.4. *Charles Darwin (1809–82).*

In Darwin's day, Sir Richard Owen (1804–92), a comparative anatomist and the first curator of the

British Museum (Natural History), championed a neoplatonic version of structuralism. Owen believed that homologies shared among all vertebrates represented a theoretical design plan in the mind of God, a common pattern that God used to create animals. Darwin transformed this ideal plan into a real ancestor. With this concept of the common ancestor, Darwin could also explain the typological data uncovered by comparative anatomists, in addition to the functionalistic evidence of adaptation that he explained by natural selection. The dualistic approach that Darwin took ensured his success. Rather than resolving the structuralist/functionalist debate, he embraced the apparent contradiction in a single, simple theory.

On the subject of the earliest origins of life itself, Darwin equivocated. In the final chapter of *Origin,* he wrote,[12] "Analogy would lead me one step further, namely, to the belief that all animals and plants have descended from some one prototype." Here we find the first mention of what was to come: the universal tree of life on which all species exist as twigs or leaves connected by their ancestors to the whole of the tree. No species consists of immutable essences nor has an independent creation. At the same time, the "chain of being" as an arrangement of extant species is also discarded. The progression from simple to complex in Darwin's tree occurred over time, leaving behind branches that would evolve into today's simpler organisms. Despite the instinctive appeal of the "chain of being" to early evolutionists, progress in Darwin's view is not linear.

With his tree of life, Darwin effectively unseated Aristotle. Because of the historical alliance of Aristotle and Christian theology in the scholastic tradition, the creation account of Genesis also fell by the wayside as a serious explanation of the origin of life. In the United States, acceptance of Darwinism among Christians came swiftly, thanks to the influence of George Frederick Wright and Asa Gray. Together, they created and advocated what we would call today **theistic evolution**, the view that Darwinism was merely God's method of creation.[13]

1.4.6. Frank Lewis Marsh (1899–1992)

Needless to say, Darwin's account of the origin of life significantly conflicts with the plain reading of Genesis. As the influence of Darwin's theory expanded, opposition arose from pastors and politicians, offering little more than a return to pre-Darwinian biology dominated by the conflict between functionalism and structuralism. A few tried to incorporate bits and pieces of transformism into the biblical creation story. Writing shortly after the Scopes trial, Byron Nelson notes that "a single pair of *dogs* likely was created, from which have come all the 40 or 50 varieties which can be seen in any large dog show."[14] Finally, in 1941, Frank Lewis Marsh produced a distinctively different vision of a biblically consistent model of biology.

Figure 1.5. *Frank Lewis Marsh (1899–1992).*

Marsh began his career as a nurse, later taking teaching positions at Adventist colleges. He earned a M.S. in zoology from Northwestern University and a Ph.D. in botany from the University of Nebraska. As a professor at Union College in Lincoln, Nebraska, he completed his first book, *Fundamental Biology,* in 1941. In it, he presented a view of biology heavily steeped in the Adventist writings of Ellen G. White. In 1944, he published his most influential work *Evolution, Creation, and Science,* an expanded version of *Fundamental Biology.* He returned to his undergraduate alma mater, Emmanuel Missionary College (now Andrews University), in Berrien Springs, Michigan, in 1960 to assume a professorship in biology. He assisted in the organization of two significant creationist groups, the Geoscience Research Institute in 1958 and the Creation Research Society in 1963.[15]

The heart of Marsh's view of biology was the **baramin**, a term he coined in *Fundamental Biology.*[16] Rather than insisting that species must be separately created, Marsh believed that the "kinds" mentioned in Genesis 1:11–12, 21, 24–25 contained many modern species. Instead of calling these "kinds" species, Marsh merged the Hebrew words for "created kinds" into *baramin.* Modern species descended of the originally created baramins. Marsh advocated using interspecific hybridization as the test for determining which species belonged to the same baramin. He wrote,[17] "Any member of a kind could cross and the product (hybrid) would also be according to the kind to which its parents belonged."

By accepting speciation, Marsh broke with many traditional interpretations and openly criticized creationists and anti-evolutionists of the nineteenth and early twentieth centuries. *Evolution, Creation, and Science* has a long chapter rejecting the scholastic tradition, and Marsh wrote,[18] "The doctrine of special creation does not teach that nature is static." Despite this decidedly anti-Aristotelian perspective, Marsh accepted baraminic fixity as a biblical principle,[19] but he gave wide latitude to the possibility of variation within the baramin. For example, Marsh accepted the actual relationship of fossil horses from *Hyracotherium* to *Equus* by proposing that they may have been members of a single baramin, in contrast to anti-evolution critics prior to him.[20]

By jettisoning the concept of species stasis, Marsh incorporated evidence for limited evolution into his creation model, though he insisted that evidence of evolution between kinds did not exist. With regard to the homologous similarities that led Darwin to propose the common ancestor of living things, Marsh wrote,[21]

> "What possible reason would a Creator have for forming them in such a way that the embryonic man and the adult sea squirt would both have notochords?" To this the creationist replies, "Who is man to attempt to assume why the Supreme Intelligence did this or that?" The evolutionist's question is absurd.

Consequently, large-scale evidence for evolution, including geographical distributions and much of homology, remained outside his explanatory sphere and within the realm of either chance or God's impenetrable will.

Despite weaknesses, Marsh's work provided an important conceptual advance in that he explicitly allowed the Scripture to shape his theories. Part of his rejection of species stasis stems from his inability to locate such a teaching in the Bible. He also quotes extensively from Scripture in *Evolution, Creation, and Science* when justifying his baramin concept. Even the Hebrew origin of the word *baramin* marks a departure from the standard of Latin and Greek in taxonomic language. Unlike so many modern creationists who omit references to the Bible out of political motivations, Marsh freely integrated the truth of the Bible with his science.

After formulating and articulating the baramin concept in *Evolution, Creation, and Science,* Marsh spent the rest of his career promoting his ideas to his fellow creationists and Adventists. He produced sixteen different books and hundreds of articles. Five years after his retirement, he published his last book, *Variation and Fixity in Nature,* a comprehensive statement of his baramin model that structurally parallels his earlier work, *Evolution, Creation, and Science*. It is from *Variation and Fixity in Nature* that a new generation began to learn of Frank Marsh's work and its importance to creationism.

1.4.7. Modern Creation Systematics

In spite of fifty years of unceasing promotion of baramins by Marsh, creationists were slow to adopt his model. Whitcomb and Morris cite his work in *The Genesis Flood,*[22] but only in 1974 did the first creationist study of a baramin appear. Hilbert Siegler published a review of hybridization in the family Canidae, noting as John Hunter did before him that dogs, wolves, and jackals all interbreed, placing them in the same Marshian baramin. In this pioneering paper, Siegler asks, "Is it not time for wide acceptance and use of" baramin?[23] Following Marsh's lead in allowing the Bible to inform science, Arthur Jones attempted to count the number of baramins on the ark by examining the biblical text and by trying to estimate the taxonomic limits of the baramin.[24]

The 1990s saw a significant revival of Marsh's ideas. The German creationist group Word and Knowledge produced an edited volume of papers that applied Marsh's hybridization criterion to various groups of plants and animals.[25] ReMine formalized the terminology of creationist systematics with his **discontinuity systematics**,[26] and Wise blended discontinuity systematics with the biblical creation account to produce the formal discipline of baraminology.[27] In 1996 a group of graduate students, Christian college professors, and amateur biologists formed an informal affiliation for collaboration on baraminology. Calling themselves the Baraminology Study Group (BSG), they have been the primary motivation behind the recent revival of interest in baraminology.[28] Members of the BSG have produced seven baraminology studies since 1997, in addition to sponsoring and organizing two conferences on creation systematics.[29] With more scientists showing interest in baraminology, the future development of creation biology looks better all the time.

1.5. Chapter Summary

Baraminology is a creationist method of biosystematics, by which organismal diversity is named and classified. In addition to endorsing the power of the word, the Bible also lays the foundation for proper biosystematics. Biosystematics finds its origins in Adam's naming the birds and beasts. From that story, we find that biosystematics is a valuable, natural, intuitive human activity. The act of discovery brings about a desire to worship and honor the Creator, thus giving value to biosystematics. Because Adam named animals without any recorded tutelage, we may infer that his naming was natural and intuitive. The act of naming organisms is a way to know the Creator more fully.

Modern biosystematics has its roots in Aristotle's essentialism and "chain of being" concept. According to Aristotle, our ability to recognize groups of organisms comes from our perception of certain essential and constant features present in all members of the group. Through scholasticism, Aristotelian views came to be associated with church doctrine, and subsequently with scientific ideas proposed by Christians. Consequently, Linnaeus (the father of taxonomy) believed in a logical order to creation and in the fixity of species. Modern scientists continue to use the binomial nomenclature he established. The turn of the nineteenth century was a time of upheaval in biology as transformism became increasingly popular. Structuralists like Geoffroy proposed that the similarity of anatomical form indicated some kind of inherent relationship between different species. Functionalists like Cuvier often continued to maintain that species fixity was the correct view. Darwin largely resolved this conflict with his theory of evolution by natural selection, which embraced both functionalism and structuralism.

In 1941, Frank Marsh produced a distinctively different vision of a biblically consistent model of biology. The heart of Marsh's view of biology was the baramin. Marsh advocated using interspecific hybridization as the test for determining which species belonged to the same baramin. Today, baraminologists carry on Marsh's pioneering work.

Review Questions

1. Give a brief definition of baraminology.
2. Explain how taxonomy differs from baraminology.
3. Give one reason why naming is so important to biology.
4. Define biosystematics.
5. Does the conventional classification of biologists follow specific rules? Should it?
6. Using a flow chart, create a brief history of the classification of organisms. Include important people and conceptual advances.
7. List the four principles of systematics derived from the Bible.
8. Give a scriptural justification for biosystematics.
9. Define scholasticism.
10. How did scholasticism influence modern biological classification?
11. Give two reasons why Linnaeus is so important to modern systematics and classification.
12. How did Lamarck contribute to the changing view of biological classification?
13. How did the structuralists contribute to Darwin's view of evolution?
14. Why was Marsh's view of biology so different?
15. Give Marsh's definition of baramin.
16. Define interspecific hybridization.
17. How did Marsh use interspecific hybridization in classification?
18. In contrast to many modern creationists, how did Marsh handle the Bible?
19. Define discontinuity systematics.
20. Who started the Baraminology Study Group?

For Further Discussion

1. If Aristotle's works had been lost to modern culture, how might biology have developed differently?
2. Would the theory of evolution have originated without Aristotle's influence?
3. Can we fit speciation into a biblical view of biology? Why or why not?
4. Explain Aristotle's influence on both creationism and evolution.
5. How did Adam name the birds and beasts?
6. Should we continue to use binomial nomenclature in baraminology? Why or why not?
7. Would the theory of evolution have originated without Darwin? Give evidence to support your answer.
8. What are some weaknesses in modern creation biology? How can these weaknesses be turned into strengths?
9. Why is discontinuity important to creation biology?
10. Should scientists spend time and effort categorizing every insignificant creature? How does one determine the significance of a creature?

CHAPTER 2

The Pattern of Life

2.1. Introduction

When we look at organisms, for example at a zoo, we do not see just a random jumble of creatures. If we look carefully, we can detect similarity among organisms, and more observation leads us to conclude that the similarity itself forms a pattern. For example, we recognize that a goose is more similar to a duck than to an eagle. The similarity relationships can extend even further. Normally we might not think of a gerbil being similar to a cat, but compared to a jellyfish, the gerbil and cat are very similar. Likewise, a monkey and a trout share much more in common than a monkey and a geranium. These similarities reveal to us that organisms were not created randomly. Instead, there is a pattern to life.

Biologists and philosophers have struggled over the nature and meaning of the pattern of life for millennia. As we explained in the previous chapter, scientists like Owen believed that the pattern revealed God's creative plan. Darwin and his followers believed that the pattern represented a great, evolutionary tree. Today, the tree still dominates evolutionary thought, but there are important attempts to change the nature of the tree. Some paleontologists have suggested that the branches of the tree shoot out very rapidly rather than gradually as Darwin imagined. Modern genomics experts are beginning to propose that the base of the tree isn't a tree at all but more like a tangled knot of roots. All of these ideas are attempts to explain the pattern of similarities that we observe in living things.

For creationists to address the pattern of life, we must first distance ourselves from our preconceived biases. Instead of thinking in terms of one pattern or another, we must reevaluate the data itself in order to discern the pattern of life. To do so, baraminologists recommend using a theoretical concept called **biological character space**. If we imagine organisms that have only two properties, length and mass, we could graph these properties on a two-dimensional surface. Thus, each organism would be represented by a single point on the graph. We know that real organisms have many more properties than just two. If we want to describe real organisms in the same way, we would have to plot a multidimensional graph,

Figure 2.1. *We readily recognize similarity between ducks, swans, and geese (all members of family Anatidae). Owls are recognizably different.*

with each dimension representing a particular property, such as color, height, or width. This multidimensional graph, wherein organisms are represented by points, is biological character space (in systematics, properties are called characters).

Returning to our imaginary organisms with only two characters, if two organisms had similar length and mass, they would be adjacent on the two-dimensional character graph. This also holds true for biological character space. Organisms like the duck and swan would end up very close in character space, with a noticeable distance between them and the eagle. In the same way, the distance between the points representing the monkey and the trout would be much less than the distance between the points representing the monkey and a geranium. Biological character space is just a formal way of describing biological similarity that we intuitively recognize.

Continuing with the idea of biological character space, the pattern of life would be the appearance of character space if we could graph every organism that ever lived. As noted above, evolutionists claim that the pattern of life would look like a tree. For example, even though the monkey and geranium appear very different today, evolutionists believe that we could trace an unbroken line of similar organisms through character space that would represent an actual lineage. Alternatively, creationists assert a fundamental "brokenness" to the pattern of life. Instead of a smooth, uninterrupted series of organisms, creationists claim that organisms form separate "lumps" in character space. We call these lumps *baramins*.

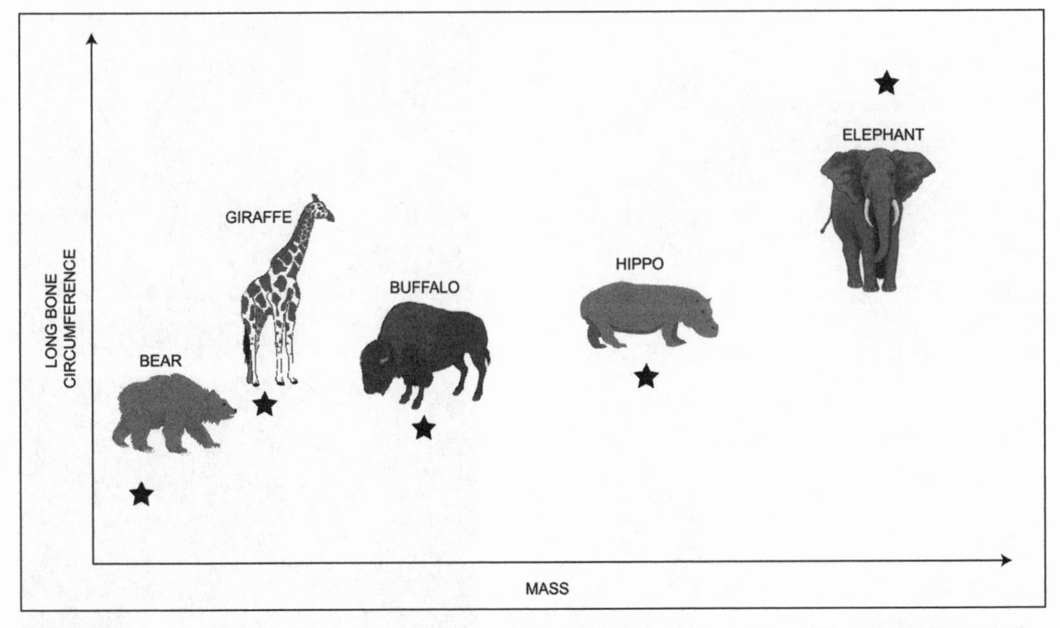

Figure 2.2. A two-dimensional slice through the pattern of life. Here we see the relationship between body mass and the circumference of long bones (humerous and femur). (Based on data in R. M. Alexander, Dynamics of Dinosaurs & Other Extinct Giants *[New York: Columbia University Press, 1989], p. 23.)*

2.2. Creationists and the Baramin

As we saw in the last chapter, Marsh accepted a certain amount of **continuity** in character space, but that continuity connected organisms which belonged to the same baramin. In devising and arguing for his baramin, Marsh claimed repeatedly that **discontinuity** was an overwhelming theme in the pattern of life.[1] By discontinuity, Marsh meant that a continuous evolutionary lineage connecting baramins did not exist. In modern terminology, we would say that Marsh believed that organisms formed discrete groups (baramins) in biological character space. To make his interpretation of character space practical, Marsh advocated interspecific hybridization as the single, defining characteristic of the baramin.

Marsh changed his ideas only slightly during his lifetime. Originally, he claimed that the hybridization criterion came from the creation account in Genesis. Marsh originally linked the phrase "after its kind" with the reproduction of organisms. A closer examination of the passage reveals that "after its kind" describes creation, not reproduction, and when God commanded organisms to "be fruitful and multiply," kinds were not mentioned. Later in his life, Marsh stopped using the Bible to argue for his hybridization criterion, but he never abandoned hybridization as the defining trait of baramins.[2]

As we noted in chapter 1, the 1990s saw a revival of Marsh's idea among creationists. Siegfried Scherer and his colleagues in Europe devised **basic type biology**, building directly on Marsh's hybridization criterion.[3] According to Scherer, two organisms belong to the same basic type (baramin) if they are able to hybridize or to hybridize with the same third organism.[4] In the United States, ReMine and Wise proposed modifications to Marsh's system that would allow a broader application of the baramin. They both noted that hybridization provides a poor basis for building a biosystematics method, as we will discuss in chapter 7. Most obviously, there are many organisms that do not reproduce sexually and so could never be classified in a baramin or basic type.

Instead of hybridization, ReMine and Wise advocated a method of approximation based on phylogenetic baramin.[5] ReMine introduced three important terms to describe groups of organisms. A **monobaramin** is a group of organisms that share a common ancestor. An **apobaramin** is a group of organisms that do not share an ancestor with any other organisms. A **holobaramin** is a group of organisms that share a common ancestor and do not share an ancestor with any other organisms. Baraminologists often equate the holobaramin with Marsh's baramin (although technically they are different). We can think of the monobaramin as a piece of the holobaramin, and the apobaramin as a group of holobaramins.

Why would we need to study pieces or groups of holobaramins? The answer to that question leads us to their most ingenious contribution to creationist systematics. Instead of trying to define the whole baramin in one shot, ReMine proposed that baramins be defined approximately with the explicit possibility of future refinement. Thus, we might find that two organisms, like the duck and goose, are very similar and probably share a common ancestor. Instead of proclaiming a duck/goose baramin, we can simply say that the duck and goose belong to a monobaramin, leaving open the possibility that other species like the swan might also belong to the same baramin.

At the same time, we might look at the creation account and see that birds are created separately from other organisms. From that, we could conclude that birds do not share an ancestor with other organisms, or in ReMine's terminology, that birds form an apobaramin. Since the creation account does not tell us about the relationships of modern bird species, we cannot claim that all birds share a common ancestor. As evidence accumulates, however, we can begin to divide the bird apobaramin into smaller apobaramins. At the same time, we can build up monobaramins by adding species to larger groups.

By continually refining our approximations of the baramin, we hope to come to a point of overlap, in which the membership of the monobaramin is the same as the membership of the apobaramin. Remember that a monobaramin is simply a group of organisms that share a common ancestor. Since the holobaramin also shares a common ancestor, the holobaramin is a special case of a monobaramin. Likewise, the holobaramin shares no ancestry with other organisms, making it a special case of an apobaramin. Thus, by **successive approximation** of monobaramins and apobaramins, the holobaramin can be detected.

Wise combined ReMine's terminology and methodology with biblical considerations to introduce the formal discipline of baraminology. In doing so, he recognized that ReMine's phylogenetic definitions, while an improvement on Marsh's baramin, still had significant limitations. For example, if we could examine organisms on the day they were created, each would be its own holobaramin, since they share no ancestry with any other organisms. Imagine if God created two identical bacteria. Since bacteria reproduce asexually, all the progeny of these bacteria would belong to two separate holobaramins, even though their ancestors are indistinguishable. Thus, although ReMine and Wise came very close to developing a workable creationist biosystematics method, more theoretical work on the nature of the baramin was necessary.

2.3. The Refined Baramin Concept

As we mentioned in the previous chapter, researchers interested in baraminology formed the Baraminology Study Group (BSG) in 1996. As various BSG members worked together on new methods for discovering monobaramins and apobaramins, the need for a new baramin concept became apparent. At the same time, the BSG recognized the value of the successive approximation method and wanted to preserve it as much as possible. Recently, they published a "refined" baramin concept, reflecting their desire to maintain the system in place but to modify some of the theoretical foundations.[6]

The development of the **refined baramin concept** began with practical considerations of how baraminologists actually do baraminology. It was quickly noted that defining the baraminology terms based on ancestry adds an unnecessary level of inference to baraminology. To assign two species to a monobaramin, first a biologist would evaluate evidence of similarity. From that evidence, ancestry would be inferred, then their assignment to a monobaramin would be inferred. At this point, the creationist should wonder how to be sure of the common ancestry inference, in order to be sure of the monobaramin inference. Since all baraminology methods rely on similarity anyway, it would simplify the process to define the baraminology terms based directly on similarity.

The refinement of the refined baramin concept replaces ancestry inference with measurement of similarity, beginning with the concepts of continuity and discontinuity. Continuity is defined simply as significant, holistic similarity, and discontinuity is significant, holistic difference. As you might imagine, it is possible to detect significant differences between any organisms if we examine particular traits. For example, it is possible to detect significant differences between the DNA of two people, possibly exonerating a person accused of a crime. To avoid these problems, we require that the similarity or difference be holistic. That is, we define continuity or discontinuity as significant similarity and difference within the whole of character space. Practically, of course, we must approximate character space by selecting a wide range of characters, including morphological, molecular, and even ecological.

With this first refinement in mind, we can now introduce the refined definitions of the baraminology terms. A *monobaramin* is a group of known organisms (or species) in which each individual shares continuity with at least one other member of the group. An *apobaramin* is a group of known organisms (or species) discontinuous with all other organisms. A *holobaramin* is a group of known organisms (or species) discontinuous with all other organisms and within which each individual shares continuity with at least one other member of the group. We can see that these definitions retain the spirit of ReMine's terminology and Marsh's original baramin.

By retaining *monobaramin* and *apobaramin,* we can still identify baraminic groups through the successive approximation method. We call evidence of continuity **additive evidence** because it provides a basis of adding organisms to a group. In like manner, we describe evidence of discontinuity as **subtractive evidence** because it subtracts organisms out of a group. By discovering and evaluating additive and subtractive evidences, we can still construct monobaramins and apobaramins and approach the holobaramin. Thus, the method-

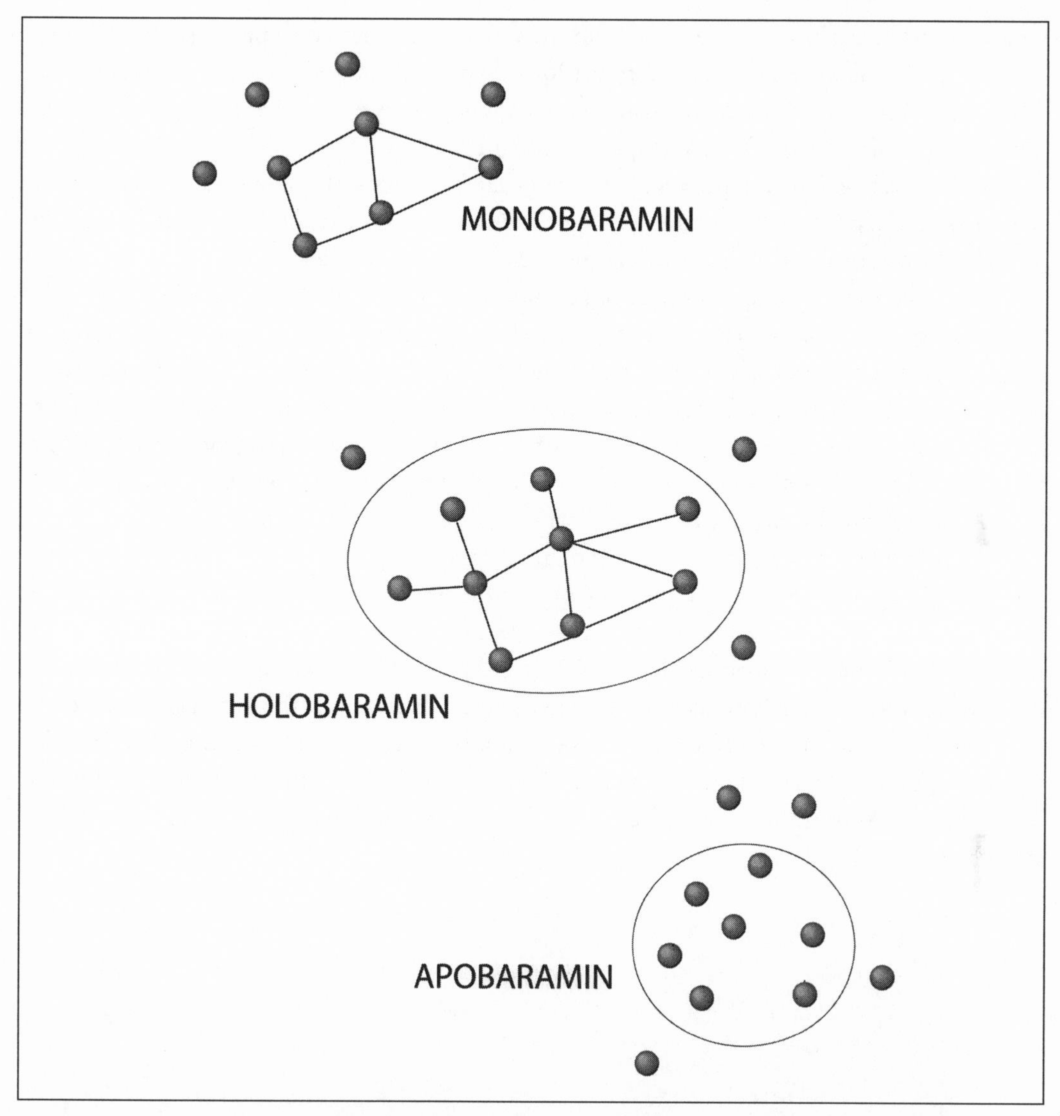

Figure 2.3. *A monobaramin (top) is a group of organisms (represented by dots) for which we can establish continuity (represented by connecting lines). An apobaramin (bottom) is a group of organisms separated from all others by discontinuity (represented by a circle). A holobaramin (middle) is a group of organisms that is continuous within the group but discontinuous with all other organisms.*

ology of baraminology as it is now practiced does not significantly change under the refined baramin concept.

The refined baramin concept also contains two more theoretical concepts that build on the notion of biological character space. As we imagine character space, we can easily envision points (representing organisms) anywhere in character space. Some creationists have proposed that this idea is incorrect. Instead, they suggest that God created character space as discrete areas where organisms are possible and bordered by areas where organisms are not possible. In other words, there are some combination of characteristics that could *not* produce a real organism. Within the refined baramin concept, these areas of possible organisms are called **potentiality regions**. Within potentiality regions, any combination of characters could produce a real organism, but between potentiality regions, no combination of characters could produce a real organism.

Potentiality regions are presently still very much a philosophical construct that help us to understand and justify discontinuity. For example, we might explain significant, holistic difference between two sets of organisms as just a lack of data. In other words, we could at some future date discover an organism that bridges the gap between the two groups. This argument could be extended indefinitely, to encompass all living things. For example, the evolutionist claims that future fossil discoveries will continue to corroborate the continuity between humans and apes. The idea of potentiality regions lends a reality to the discontinuity. If there are areas of character space (combinations of characteristics) where organisms *cannot* exist, then discontinuity is not merely a lack of knowledge. The difference between two organisms in separate potentiality regions is real and cannot be bridged with more knowledge.

All of the organisms that exist in a potentiality region at any point or period of history are defined as a baramin. The baramin can be thought of as the physical manifestation of the theoretical potentiality region. Whereas the holobaramin technically applies only to organisms for which we have evidence of their existence, the more theoretical baramin can include organisms that left no trace of their existence. For example, the organisms on the ark or at creation could be described as members of a baramin but not as members of a holobaramin (unless by some chance we found definitive evidence of their existence). With the baramin, we can begin to theorize about the history of organisms since creation or during the Flood.

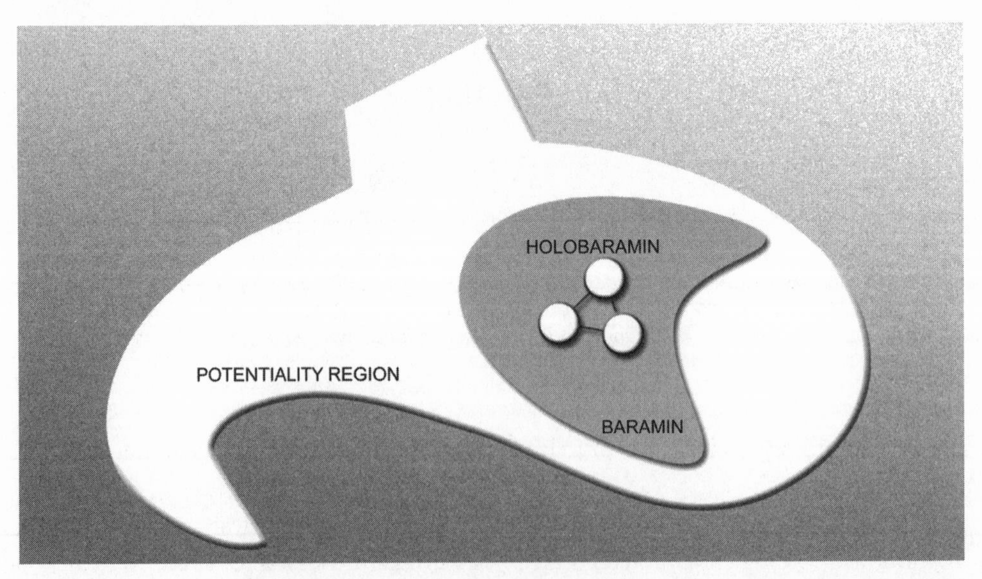

Figure 2.4. *The potentiality region marks an area of biological character space where organisms are possible. The baramin is the actual group of organisms that occupy that potentiality region at any point or period in history. The holobaramin is composed of members of a baramin of which we actually have evidence.*

2.4. *Justifying the Refined Baramin Concept*

Before we discuss the practical issues of baraminology (the topic of the rest of the book), it is important to try to justify this view of the pattern of life. To do so, we will review biblical, theological, and biological evidence for the refined baramin concept. Unlike Marsh's early view of "after its kind," we will claim no unequivocal justification from any single source. Instead, we will discuss a number of evidences that seem to depict life as discrete units. Taken together then, we can construct a holistic justification of the refined baramin concept. Since holism is always preferable to overemphasizing one potentially wrong biblical interpretation, the justification of the refined baramin concept is actually stronger than Marsh's biblical justification of the baramin.

2.4.1. Biblical Considerations

Following Marsh's example, creationists often look to the term *min* (Hebrew for "kind") to justify the baramin, sometimes even claiming that *min* is an absolute category of biological classification. Unfortunately, there is little evidence for this position. The word *min* appears thirty-one times in the Hebrew Old Testament, thirty in the Pentateuch and once in Ezekiel. The usage of *min* in the Pentateuch is concentrated in the creation account (Gen. 1:11–12, 21, 24–25) and the dietary laws (Lev. 11; Deut. 14). All biblical instances refer to *min* of animals or plants, but the exact meaning of the word is unclear. Occasionally *min* appears in extrabiblical writings, and some scholars argue that these usages indicate a root meaning of "division" for *min*. Since the precise meaning of *min* is still not clear, it is best to avoid using *min* as the foundation of the baramin.

Instead of giving absolute support of the refined baramin concept, the Scripture provides much indirect evidence for both continuity and discontinuity. First, as we already discussed in chapter 1, Adam's naming of the birds and beasts in a single afternoon becomes feasible if he named groups larger than species. Second, baramins help to explain the capacity of the ark. If Noah took only two of every (unclean) baramin on the ark, there would be sufficient room for the people, the animals, and plenty of provisions.

Evidence for widespread discontinuity also comes from the creation account. God created plant life, swimming things, flying things, land animals, and humans *separately,* on three different days of Creation Week. These separate acts of Creation imply fundamental discontinuities between the groups, and their creation (and completion) on separate days of creation reinforces the discontinuity between them.

Finally, we may also infer evidence of discontinuity from biblical chronology. The genealogies of the Bible imply that creation occurred only six thousand years ago. A six-thousand-year history would seem to preclude large-scale evolution. Although we cannot be dogmatic about this point, six millennia seems to be insufficient time to evolve completely new phyla. The origin of new species could certainly occur in such a short time span, but large-scale evolution is probably limited, if it occurs at all. Thus the baramin is probably between the size of the phylum and species.

2.4.2. *Theological Considerations*

To understand continuity and discontinuity from a theological perspective, we must first discern God's motivation for creating. It may sound sacrilegious to attempt to ascribe motivation to God, but the Bible gives us fairly clear insight into God's desires. First, the Bible teaches that God desires to be known by us (John 1:14; Matt. 27:51; Rev. 21:3; Gen. 3:8–9). We can even view the Bible itself as a manifestation of this desire. Thus, we can understand God simply because He reveals Himself to us in Scripture.

Most importantly for our present discussion, creation itself appears to be a manifestation of God's desire to be known (Ps. 19:1–4; Rom. 1:19–20). Creation is not simply a habitation for humans but a revelation of God and His attributes. Like the written revelation in the Bible (Isa. 40:8), it is reasonable to expect that God's revelation in creation would somehow survive or persist. A persistent revelation in creation has two important biological consequences.

First, if we believe that the pattern of life formed by individual organisms contains information about God, the pattern must not be so variable as to obscure the information in it. In other words, suppose God created a particular pattern of organisms in character space in order to reveal His own nature. Without any constraints on the variability of organisms, changes could occur in six thousand years that would alter or perhaps even abolish the original information-containing pattern. As a result, we may conclude that God created organisms to be stable over time, preventing loss of the information in the pattern of life.

At this point, it might be easy to conclude that God created organisms as fixed species, in the essentialist sense. Species fixity seems to perfectly preserve the pattern of life with no degradation of information. In actuality, fixity would degrade the information in the pattern of life just as much as unlimited variability would. Species fixity only works if *all* changes, including environmental changes, are disallowed. A species perfectly adapted to its environment would go extinct in response to drastic environmental change. In fact, adaptability is the only way for revelation to persist in the pattern of life. Organisms must be created with the potential for variation in order for the pattern to survive environmental changes. Taken together then, the persistence of revelation in the pattern of life implies adaptation and variation (and consequently continuity) within definite limits (discontinuity).

2.4.3. *Biological Considerations*

Because the rest of the book will present numerous examples of continuity and discontinuity among species, we will limit our discussion here to the question of discontinuity. Within creation biology, we need merely look with Linnaeus to interspecific hybrids for evidence of continuity between species. In modern evolutionary biology, phylogenetic relationship among all organisms is the assumption. The concept of discontinuity in the evolutionary tree of life is far more important for justifying the baraminology approach.

Before addressing the reality of discontinuity, we must remember that we need not demonstrate discontinuity at the level of the baramin in order to justify looking for discontinuity at that level. The refined baramin concept merely serves as a basis for investigating and understanding organisms. Showing discontinuity at the level of the baramin is a goal of the application of the refined baramin concept. To justify the concept, we need merely provide evidence that some discontinuity might exist. Such evidence warrants the search for more discontinuity.

Since discontinuity is a holistic difference, any evidence of continuity between all organisms must necessarily be reductionist (non-holistic). As we look to these evidences, we find that indeed they are very reductionist. Evidence of universal continuity includes DNA, RNA, and protein as the primary biomolecules; the metabolism of DNA (replication), RNA (transcription), and protein (replication); the common genetic code; homochirality of proteins; lipid bilayer membranes; and energy conversion through electron transport. All organisms possess these traits, which are counted as evidence of derivation from a single ancestor.

When examined more closely, however, these evidences of universal ancestry display important differences in different organisms. For example, the "universal" genetic code is not universal to all organisms. The National Center for Biotechnology Information lists fifteen variant codes from a variety of organisms, including mycoplasma bacteria, some species of the yeast genus *Candida,* and some sea squirts. These variant codes resemble the "universal" code, but as many as six nucleotide codons code for alternative amino acids.

The similarity of the variant codes to the "universal" code would seem to argue for a derivation of the variant from the universal. The details of such a transformation present significant challenges to this interpretation. Alteration of the genetic code would require simultaneous alteration of all of the protein-coding genes in the organism. If the code changed first, the organism's proteins would be mistranslated. If the codons of genes changed first, the organism would still mistranslate the proteins using the old genetic code. Thus, variant codes are not readily derived from the "universal" code. Because variant codes impact the entire organism through the translation of all proteins, variant codes count as a significant, holistic difference–a discontinuity. As we noted above, the existence of some discontinuity justifies the search for other discontinuity, by application of the refined baramin concept.

2.5. What Is the Pattern of Life?

In this chapter, we contrasted a view of organisms as perfectly continuous with one that places real discontinuities into the pattern of life. While the question of continuity and discontinuity certainly pertains to baraminology, the existence of discontinuity does not resolve the actual pattern of life. For example, we could examine a painting and deduce that it was painted with watercolors rather than oils, but that would help us very little in resolving what the painting depicted. One advantage of a baraminic view of life is its flexibility. While ongoing research seeks to understand the pattern of life more fully, the baraminic view can encompass a variety of models and attributes of the pattern of life.

As we review historical attempts to describe the pattern of life, we will see that this history shares much in common with the old Indian fable of the blind men and the elephant. As each blind man grabbed only part of the elephant, each came away with a different impression of what the elephant was like. One claimed that the elephant was like a tree, another like a snake, another like a wall. In the same way, as biologists and philosophers look at the pattern of life, each is able to describe part of the pattern. Each description is correct in its own limited way. As more information about organisms comes to light, it is apparent that the real pattern of life is much more complex than any past efforts to describe it.

Aristotle's chain of being represents one of the earliest attempts to describe the pattern of life. He arranged organisms in a linear pattern from simple to complex, as we explained in chapter 1. Although modern biologists reject Aristotle's linear view, it still holds a strong appeal. The pattern appears to have some validity if not carried to extremes. Some organisms are indeed less complex than others. Unlike evolution, which insists upon a tree of life, the refined baramin concept accommodates the Aristotelian chain as a potential feature of the pattern of life. Because the pattern of life is largely created rather than evolved, an arrangement from simple to complex is entirely legitimate.

In like manner, a hierarchical pattern of life also appears to be a legitimate description of organisms. The hierarchy was strongly advocated by Sir Richard Owen, who interpreted it as a revelation of God's design plan. Later, Darwin infused the hierarchy with an evolutionary meaning, transforming the design plan into a genealogical tree. Although we reject the historical interpretation of the evolutionary tree, the hierarchical pattern has a degree of authenticity to it. Once again, the refined baramin concept accommodates this view of life. Because the primary emphasis of baraminology is the baramin, supra-baraminic classification and patterns remain active areas of research. A hierarchy could be a legitimate interpretation of one aspect of the pattern.

More recently, several researchers have proposed even more complex models of the pattern of life. Wise suggested that the actual pattern of life is a complex network of similarity, like a fishnet with organisms emerging at the joining of individual lines of similarity.[7] Such a view could explain why some view the pattern and see a tree, while others see a progression, while still others see no pattern at all. The complexity of the network allows for the existence of true hierarchical or even progressive patterns, each of which could be derived from the network by viewing it in a particular way. Thus, Wise's model subsumes the ideas of Aristotle, Owen, and Darwin.

Reynolds affirmed this position and offered a philosophical model of a network-like pattern of life, derived from the work of Plato.[8] According to Reynolds, the pattern of life dictates what organisms could exist, and all possible organisms were in fact created. Thus, by examining the pattern of life, the Platonic biologist could actually predict the existence of undiscovered organisms based on biological principles of form. This strongly reminds us of the evolutionary tree, which successfully predicts the existence of numerous morphological intermediates. Combined with Wise's view of the network, Reynolds's proposal fits well with the refined baramin concept and provides the basis for future research.

In addition to overarching models, many biologists have proposed interpretations of specific features of the pattern of life. These explanations range from species fixity to punctuated equilibrium, but at their core, each tries to explain particular aspects of the pattern. As we will see, the refined baramin concept can also explain these well-known attributes of the pattern.

In chapter 1, we argued that biosystematics ought to allow the interplay of human intuition. Adam's intuitive naming of the animals implies that the pattern of life should be intuitively recognizable. Conventional biosystematics overemphasizes one or two characteristics (called synapomorphies) to define biological groups, frequently leading to counter-intuitive classifications. For example, some evolutionary systematists advocate placing birds in the dinosaur group. In contrast, the baramin as a holistically defined group is much more amenable to the use of intuition.

Evolutionists often use evidence of microevolution and speciation to argue for macroevolution by extrapolation. Research in this area has revealed that the pattern of life is far from a rigid construct. Instead, the pattern appears to have a good deal of plasticity created in it. The refined baramin concept not only accommodates the malleability of species; it can explain it. As we discussed above, in order for the revelation of God in the pattern of life to persist in a changing world, baramins must be capable of adaptation and change. The refined baramin concept therefore predicts the kind of small "evolutionary" changes that we observe today.

In addition to explaining the occurrence of biological change, the refined baramin concept could also provide a positive understanding of one mechanism of change. Darwin's natural selection provided a plausible and demonstrable mechanism to explain environmental adaptation. With the refined baramin concept, natural selection can be viewed as a post-Fall stabilizing mechanism. With errors, death, and other changes possible in a post-Fall world, baramins would need a mechanism (possibly several mechanisms) to persist. Natural selection provides such a mechanism by eliminating biological deviants and preserving better-adapted individuals. Thus, rather than the primary agent of change, natural selection actually resists it in the baraminology model.

According to Linnaeus's early view, the pattern of life consisted of immutable species, all of which originated by direct creation. To infer species fixity, Linnaeus observed a stability of species through reproduction. Because the refined baramin concept advocates the persistence of the baramin, not fixity of species, it would be unfair to claim support of species stability within baraminology. Instead, we will merely note that species stability is consistent with the persistence of God's revelation, a key feature of the refined baramin concept. We will return to the question of species and baraminic stability in chapter 11.

In the first half of the twentieth century, Goldschmidt proposed a peculiar interpretation of evolution that won him little support from his peers. According to his idea of macromutation, occasionally a catastrophic genetic alteration produces a leap forward in evolution without evidence of intermediate species. His view is often parodied as predicting the hatching of a fully formed bird from a reptile egg. In reality, we might interpret Goldschmidt's view as an attempt to preserve evolution in light of intuitively obvious discontinuity. Since the existence of discontinuity is predicted within the refined baramin concept, the macromutation mechanism is rendered unnecessary.

Although frequently–and unfairly–compared to Goldschmidt's macromutation model, Gould and Eldredge's punctuated equilibrium actually derives from an application of population genetics. In promoting their theory, however, they have emphasized the quantum nature of biological similarity. Instead of species continuously grading one into another, we actually observe discrete species (and higher categories) with comparatively little evidence of transition. Gould and Eldredge interpret this pattern as evidence of rapid evolution in small populations, wherein evidence of transition is not preserved. Alternatively, the refined baramin concept actually predicts the kind of discontinuity Gould and Eldredge emphasize.

The refined baramin concept also subsumes previous techniques of creationist biosystematics. In particular, the practicality of Marsh's and Scherer's views of the baramin can be explained with the refined baramin concept. Both emphasize the value of interspecific hybridization as a baraminic membership criterion, but as we have seen, neither has a good creationist justification for why hybridization should reveal biosystematic patterns. With the refined baramin concept, interspecific hybridization reveals significant, holistic similarity (continuity). As we will discuss more fully in chapter 7, the production of a hybrid indicates a fundamental similarity of genetics, cell biology, development, and anatomy between the parent species.

2.6. Chapter Summary

Biologists and philosophers have struggled over the nature and meaning of the pattern of life for millennia. In order to analyze patterns of life accurately, creationists must first learn to step outside of their preconceived ideas about the pattern. The theoretical concept of biological character space aids the evaluation of such patterns.

Marsh accepted a certain amount of continuity in character space, but continuity only connected organisms that belonged to the same baramin. In devising and arguing for his baramin, Marsh claimed repeatedly that discontinuity was an overwhelming theme in the pattern of life. Marsh also advocated interspecific hybridization as the single, defining characteristic of the baramin. In the 1990s, Siegfried Scherer began building upon Marsh's hybridization criterion, devising basic type biology. According to Scherer, two organisms belong to the same basic type if they are able to hybridize or to hybridize with the same third organism. ReMine and Wise also proposed modifications to Marsh's system to expand it to include organisms that could not be analyzed by hybridization (such as asexual or fossil organisms).

Researchers of the Baraminology Study Group published a "refined" baramin concept, reflecting their desire to maintain the terminology and methodology of Wise and ReMine but to modify some of the theoretical foundations. Based on biblical, theological, and biological considerations, the BSG members introduced the following definitions:

- *continuity/discontinuity,* significant, holistic similarity or difference;
- *monobaramin,* a group of known organisms (or species) in which each individual shares continuity with at least one other member of the group;
- *apobaramin,* a group of known organisms (or species) discontinuous with all other organisms;
- *holobaramin,* a group of known organisms (or species) discontinuous with all other organisms and within which each individual shares continuity with at least one other member of the group;
- *potentiality region,* an area of biological character space in which organisms are possible, surrounded by areas where organisms cannot exist; and
- *baramin,* the actual organisms found in a potentiality region at any point or period of history.

The pattern of life is an attempt to explain God's design. As we view the attempts of others to describe the pattern, we find that the refined baramin concept can accommodate them very well.

Review Questions

1. What is biological character space?
2. Why use biological character space when analyzing the pattern of life?
3. What did Marsh mean by the term *discontinuity*?
4. How did Marsh change the way in which he argued for hybridization?
5. Describe basic type biology.
6. Define *monobaramin, apobaramin,* and *holobaramin.*
7. What purpose do the terms *monobaramin, apobaramin,* and *holobaramin* serve in the practice of baraminology?
8. What is successive approximation?
9. Why was a refinement of Marsh's ideas necessary?
10. What is subtractive evidence?
11. What is additive evidence?
12. How are additive and subtractive evidence used to identify baramins?
13. Describe potentiality regions. What benefit comes from these theoretical regions?
14. List three important biblical considerations that support the refined baramin concept.
15. How does the ark help us to understand the baramins?
16. Define pattern of life.
17. Give a biological evidence of discontinuity.
18. Why does species fixity not describe the whole pattern of life?
19. Describe Goldschmidt's concept of macromutation.
20. Describe Gould and Eldredge's punctuated equilibrium.

For Further Discussion

1. Was Marsh's explanation of the baramin sufficient for developing a new classification system? Why or why not?
2. Is it possible to disregard our biases? If so, how? If not, why?
3. Should creationists continue to use the conventional classification system? Why or why not? List the pros and cons of using the classification system, and justify your argument with biblical references.
4. What theological considerations should we take into account when building a case for baraminology?
5. Should we continue to refine the baramin concept? Why or why not?
6. How might an evolutionist argue for continuity of all living things (universal ancestry)?
7. Why do you think there has been such a struggle over understanding the pattern of life?

8. Is it possible to have a completely unbroken line of organisms through biological character space? Why or why not?
9. Why is there a fundamental "brokenness" to the pattern of life? What caused it? What biblical references can you cite to support your conclusions?
10. Would an evolutionist ever argue for a brokenness in the pattern of life? Why or why not?

CHAPTER 3

The History of Baramins

3.1. Introduction

The early chapters of Genesis present a very clear and unmistakable outline of earth history. Despite the clarity of these chapters, Christians have vigorously debated their meaning for more than a century. Rather than rekindle that debate here, we will proceed with the assumption that the Bible speaks clearly and authoritatively on all matters it mentions. We accept the historicity of the events described in Genesis 1–11 and a chronology similar to Ussher and Lightfoot. In particular, we believe that creation occurred in six consecutive days, that human and animal death arose as a result of the Fall of man, that a global Flood destroyed all terrestrial animal and human life except for what was preserved in the ark, and that the tower of Babel generated our modern diversity of human language and culture. This chapter will give an overview of this early history with emphasis on its impact on organisms.

In addition to the testimony of Scripture, recent advances in creation research have illuminated many details about these events. Theories like Catastrophic Plate Tectonics and post-Flood climate models have important implications for the history of life. As scientists, we should seek to correlate these theories to develop a comprehensive and unified model of creation. As we discuss scientific theories in this chapter, remember that theories should be accepted provisionally as probable models of reality. Although future research may require the modification or even rejection of many of these ideas, the theories discussed here are our best understanding based on modern research.

3.2. Creation

Creation occurred roughly four thousand years before Christ or six thousand years before the present day. According to Genesis 1, God completed creation in only one week. The Bible records that plants originated on Day 3, "fowl" and water-dwellers on Day 5, and "cattle," "creeping things," and "beasts" on Day 6. The origin of these recognizable groups on separate days implies fundamental discontinuities between them, but we must understand precisely which organisms are recorded in these passages to understand which organisms have a record of creation. As we read the creation account of Genesis, we must remember not to force our modern scientific perspectives on the text. The collective terms used in this passage (e.g., "fowl") should not be equated with groups used in modern scientific classification (e.g., Aves). We need to examine the Scripture and the Hebrew language carefully to understand the meaning of these terms.

The first recorded act of biological creation is the origin of vegetation. "And God said, Let the earth bring forth grass, the herb yielding seed, and the fruit tree yielding fruit . . . and it was so" (Gen. 1:11). The "grass" is the Hebrew word *dese,* which refers to green growth.[1] The "herb" is *eseb,* which also refers to green plants.[2] "Tree" is *ets,* referring to the firmness of wood. Some scholars believe that the author intended to convey three major types of plants, but the majority of scholars favors only two types of plants.[3] According to the latter view, God used *dese* as a general term for plants, with two types, *eseb* (herbaceous plants), and *ets* (woody plants). We need not believe that *eseb* and *ets* refer to discontinuous groups of baramins. It is possible that God could have created an *eseb* and an *ets* as members of the same baramin.

The creation of vegetation on Day 3 may give us clues to the creation of other unrecorded organisms. God created plants immediately after the origin of the land, but two days before the creation of any animal life. The close proximity of the origin of land and plants may imply that God views plants as *part* of the land. Following this view, it is possible that some fungi originated on Day 3, since many mushrooms also grow on the ground. Similarly, microorganisms may be part of the water or soil. If correct, many microbes originated with the water-covered earth on Day 1. This view may explain the presence of microfossils in sedimentary rocks associated with the origin of the land.[4]

Following the creation of plants, God created some animal life on Day 5. "God created great whales, and every living creature that moveth . . . and every winged fowl after his kind" (Gen. 1:21). "Whales" are *tannin,* a word frequently translated "serpent" or "dragon."[5] The *tannin* figured prominently in the pagan cosmologies at the time of Moses, and we may infer from the biblical and extrabiblical usage that *tannin* refers to large animals of the sea.[6] Whatever the *tannin* are in this passage, God created them along with all other creatures in the sea and air on Day 5.

In Leviticus 11, we find assistance for understanding the "fowl" spoken of in Genesis 1:20. The Hebrew term *op* appears to be a general term for creatures that fly.[7] The Levitical dietary law includes bats (Lev. 11:19) and insects (Lev. 11:22) in its list of clean and unclean *op*. The Bible calls a flying insect a "flying creeping thing that goeth upon all four" (Lev. 11:20). It seems probable, then, that all flying creatures were created on Day 5, including pterodactyls and *Archaeopteryx*. Since *op* refers to flying creatures, the creation of flightless birds appears to be unrecorded, unless they are grouped with the land creatures of Day 6 or arose after Creation.

"Living creatures" is a compound term found frequently in the creation account. "Living" is the Hebrew *hayya,* a word derived from the word for "life." *Hayya* also describes water in later books (Cant. 4:15; Jer. 2:13; 17:13; Zech. 14:8), revealing its broader usage to depict movement and animation.[8] The second term, "creatures," is the Hebrew *nepes,* a word

DAY OF CREATION

3RD	5TH	6TH

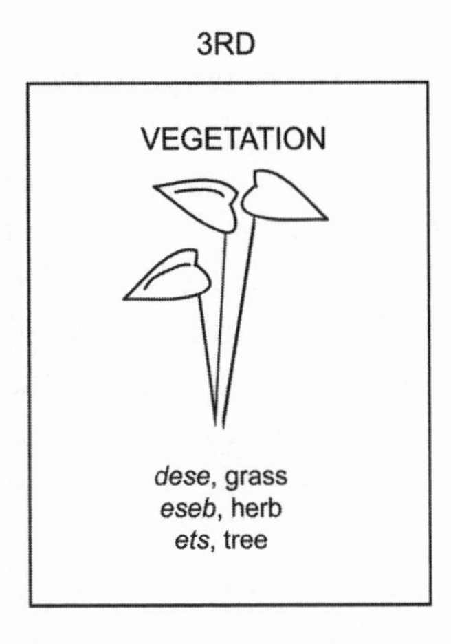

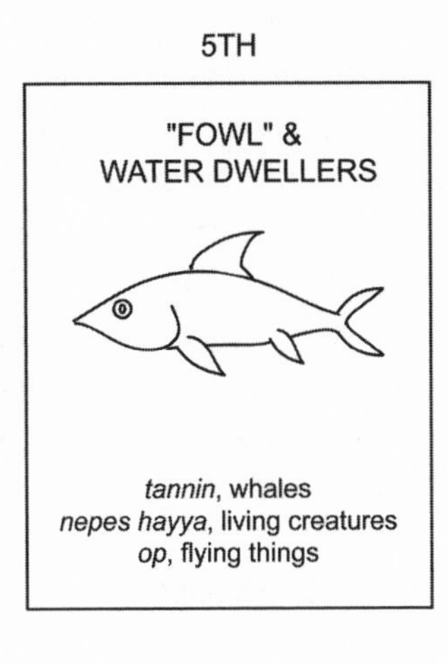

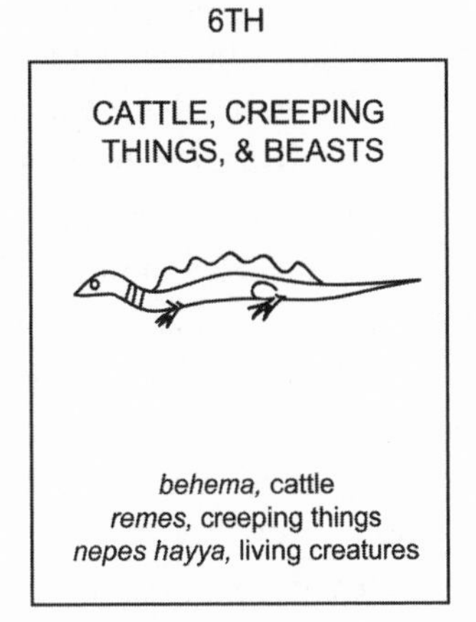

that also refers to life. Originally, *nepes* came from a term meaning "breath," but by the time Moses recorded the creation account, *nepes* had assumed its broader meaning of "living thing."[9] Together, the terms are used to describe creatures in the water (Gen. 1:20–21), creatures on the land (Gen. 1:24), and humans (Gen. 2:7, translated "living soul"). Plants are never called "living creatures," suggesting that the Bible distinguishes between the life possessed by humans and animals and the vitality of plants.

God concluded the creation events of Day 5 with a blessing and His first command to "be fruitful, and multiply, and fill the waters in the seas, and let fowl multiply in the earth" (Gen. 1:22). Curiously, the Bible also records that "waters brought forth abundantly" at the origin of the sea creatures; nevertheless, this abundance was apparently not enough to fill the oceans with life. God did not command the plants to multiply, perhaps indicating that they already covered the land as part of the land itself.

On Day 6 of Creation, God concluded His work with the creation of "cattle, and creeping thing, and beast of the earth" and human beings (Gen. 1:24–31). "Cattle" is the Hebrew *behema,* which apparently refers to four-legged creatures.[10] Although *behema* is most frequently used for domesticated animals, the word is used in Job 40:15 for a particularly spectacular *behema* that creationists have long associated with a sauropod dinosaur, such as apatosaurus.[11] Here in Genesis, we find the origin of all quadrupeds, including the dinosaurian sort, on Day 6 of Creation.

The second class of creature, the "creeping things," is a single word in the Hebrew: *remes. Remes* is derived from the root *rms,* meaning "to creep" or "to teem."[12] In the Levitical law, the noun form is identified as unclean. In two instances, *remes* refers to water-dwellers (Ps. 104:25; Hab. 1:14), but this appears to be a later and unconventional usage. In Genesis 1:24, *remes* probably refers to creatures that move along close to the ground, including reptiles, amphibians, and rodents.

The final group created was the "beast of the earth." The "beast" is a form of the same word for "living" in "living creatures," *hayya.* Thus, Genesis 1:24 may be approximated as

"God said, Let the earth bring forth the *hayya* creature . . . cattle, and creeping thing, and *hayya* of the earth." The distinction between the *behema* and "*hayya* of the earth" may in this instance be one of domestication. As noted above, *behema* most often refers to domesticated quadrupeds, while we find "*hayya* of the earth" refers primarily to wild creatures. If correct, this would imply that domestication is a divinely created relationship for certain animals.

After the creation of human beings, God repeated the command to "replenish the earth" (Gen. 1:28), again implying that the created organisms did not occupy the entirety of the available environments. Following the command to fill the earth, God commanded humans, birds, beasts, and creeping things to eat only plant life. The repetition of the word *every* in Genesis 1:30 ("every beast," "every fowl," "every thing that creepeth") rules out the possibility of an organism created to be a predator. All animals were created to be strict herbivores, including the cats, the snakes, and the "carnivorous" dinosaurs. The biblical text makes no exceptions.

Genesis 2 gives a more detailed account of the creation of man on Day 6. Unfortunately, the viewpoint and detail of chapters 1 and 2 are different, leading many to conclude that the first two chapters of Genesis contradict each other. When we consider the viewpoint of chapter 2, we find that there are no contradictions at all. Chapter 1 is a very broad view of all of creation, while chapter 2 more narrowly and personally describes the creation of human beings. Genesis 2:5 states that the "plant of the field" and the "herb of the field" had not grown. The qualification "of the field" tells us that the author intends to convey that the events about to be described took place before the beginning of agriculture (before the creation of humans).

Genesis 2:5 also states that "God had not caused it to rain upon the earth," and verse 6 further elaborates that "there went up a mist from the earth, and watered the whole face of the ground." Curiously, many creationists have interpreted the mention of rain in these two verses much more broadly to say that there was no rain before the Flood. When we remind ourselves of the limited, personal perspective of chapter 2, we find it hard to extrapolate from no rain on Day 6 in Eden to no rain anywhere on the earth until the Flood, over sixteen hundred years later. This brief mention of the mist should be understood in the context in which it is written, as speaking of the region of Eden during the Creation Week. The occurrence of tree rings in pre-Flood fossilized tree trunks confirms pre-Flood seasonal rains.[13]

In keeping with the personal intimacy of chapter 2, we find that God created Adam not by declaration but by forming him from the "dust of the ground." After receiving the "breath of life," man became a living creature (*nepes hayya,* the same term used for animals and sea creatures). The very construction of this passage serves as a polemic against human evolution. Man became *nepes hayya,* but a *nepes hayya* (living creature) did not become man.[14] Those who interpret the origin of humans as the infusion of a soul into preexisting, soul-less primates find no support for their view from the biblical account of the creation of humans.

Genesis 2:8 begins an extended description of the region of Eden in which God planted the garden that man would inhabit. The geography of Eden reminds us that the "world that then was . . . perished" (2 Pet. 3:6). No modern river splits into four rivers, yet just such a river existed in Eden (Gen. 2:10). Many modern Bible scholars try to locate Eden in our modern world based on the location of the present Tigris (Hiddekel) and Euphrates (Gen. 2:14), but the complete destruction of pre-Flood geography prevents modern discovery of the location of Eden.

3.3. Fall

The Fall of man is arguably the most significant event in the history of life on this planet, comparable to the Flood's impact in the field of geology. Many attributes of modern living things testify to the devastation of the Fall. God created organisms to be herbivorous, but today carnivory is common. God created humans to live forever, but today everyone dies. God placed humans in an idyllic garden, with access to a Tree of Life. Today, Eden and the tree are lost forever. Genesis 3 reveals some of the details of the event that changed the living creation from its original perfection to its modern condition.

The Fall came as the newly created humans partook of the forbidden fruit of the Tree of Knowledge of Good and Evil. A serpent deceived Eve into eating the fruit. She subsequently gave the fruit to Adam "with her," and the Fall of man was complete. As the serpent predicted, "their eyes were open," and they made fig-leaf aprons to cover themselves. When God came to the garden in "the cool of the day," Adam and Eve hid. God ultimately pronounced a curse upon creation, under which it still groans (Rom. 8:22). The prophet Isaiah foretold a coming day in which the curse would be lifted and predators would peacefully coexist with their prey (Isa. 11). Perhaps as a counterpoint to the deceitful snake in Genesis, Isaiah records of that coming day that "the suckling child shall play on the hole of the asp" (Isa. 11:8).

The central figure of the Fall is the enigmatic "serpent . . . more subtil than any beast of the field" (Gen. 3:1). A great deal of speculation has been written about this serpent. Liberal scholars emphasize the mythological symbolism of serpents and dragons,[15] while more literal-minded young-earth creationists infer that snakes once walked about on four legs (Gen. 3:14). Because so much has been written about the serpent, a biological analysis of its identity is certainly warranted.

The Hebrew text uses ten words to refer to snakes, vipers, and serpents. Nearly all of them pertain to one of six species of venomous serpents found in the Middle East. In Genesis 3, "serpent" is the Hebrew word *nahas,* carrying the general meaning of "snake." In six passages, *nahas* parallels words for venomous serpents. While the general usage of *nahas* refers to snakes, the word can apply broadly to other types of reptilian creatures. For example, Isaiah describes Leviathan, the dragon of the sea, as *nahas*. This wider usage of the word, although rare, admits the possibility that the *nahas* of Genesis 3 refers not to a snake but to another type of reptilian animal.[16] Based on this usage, we can conclude that the *nahas* was a created, reptilian organism, perhaps in some ways the most impressive of all God's animal creation (Gen. 3:1). At some point after God declared everything "very good" (Gen. 1:31) and before the beginning of Genesis 3, Satan fell into sin and selected the serpent as his tool for destroying God's creation.

Associating the leglessness of snakes with the Curse on the *nahas* dates to antiquity. For example, Josephus accepts this interpretation of the Curse.[17] The actual wording of the

Curse is "upon thy belly shalt thou go, and dust shalt thou eat all the days of thy life" (Gen. 3:14). Since lizards, crocodilians, and even turtles go about on their bellies, the Curse actually offers few clues to the identity of the *nahas*. The universal association of the term *nahas* with serpents, particularly venomous serpents, leaves little room for interpreting the word as anything but a reptile; however, its association with such creatures as the *tannin* and Leviathan prevents more specific identification. It may have been a legged serpent as Josephus believed, or it could have been any type of reptile relegated to a crawling gait after the curse. It could be a species extinct in our modern world, or it could even refer to an individual animal without applying to all members of the same species.

The Curse itself stems from the disobedience of Adam and Eve to God's command not to eat of the fruit of the Tree of Knowledge of Good and Evil. "And the LORD God commanded the man . . . of the tree of the knowledge of good and evil, thou shalt not eat of it: for in the day that thou eatest thereof thou shalt surely die" (Gen. 2:16–17). The promise of death raises something of a conundrum, in that Adam and Eve clearly did not die "in the day" that they ate. In fact, Adam lived 930 years before he died (Gen. 5:5).

Some believe that we should interpret "death" as spiritual separation from God in order to accommodate this apparent contradiction, but God states in the Curse, "for dust thou art, and unto dust shalt thou return" (Gen. 3:19), obviously referring to physical death. The answer to this apparent contradiction comes from a careful study of Hebrew. The Hebrew phrase "in the day" can be found in other scriptural passages as a figure of speech meaning "when" or "if." Based on this broader meaning of "in the day," we need not interpret God's promise to mean "on the exact day that you eat of the tree of knowledge, you will immediately die." Instead, the promise could be rendered "if you eat of it, you will surely die."[18]

Upon discovering man's sin, God pronounced a Curse on creation. From the few short verses that record God's Curse (Gen. 3:14–19) and from Isaiah's prophecy of a restored creation (Isa. 11), creationists have inferred the source of nearly every form of "natural evil," including noxious plants, poisons and toxins, diseases, carnivory, parasitism, aging, and all animal death. Because creationists attribute so much to the Curse, we need to examine it in detail to ensure that our understanding is based on the text or on reasonable inferences.

God curses the snake beginning with the phrase, "Thou art cursed above all cattle, and above every beast of the field" (Gen. 3:14). By implication, the cattle and beasts of the field must also be cursed, since the serpent is cursed "above" them. Unfortunately, the author does not elaborate on the nature of the general Curse on the serpent and other creatures. Since the promise of death is so prominent in the narrative of the Fall, and because of Isaiah's prophecy, we may reasonably infer that the curse shared by the animals was death. As a result, animal death would be impossible prior to the Fall. Fatal disease, fatal toxins, predation, and carnivory would have been unknown before sin entered the world.

The Curse on Eve consisted of two specific changes: sorrow in childbirth and conflict in marriage (Gen. 3:16). By implication, childbirth before the Fall would have been significantly less painful than it is now, but the condition of animal birthing before or after the Fall is not given or implied in the Curse. Additionally, in the beginning, Adam and Eve were in perfect communion, but their sin brought conflict between husband and wife. Like Adam and the animals, Eve also shared in the promise of death.

The author of Genesis 3 records a more extensive curse on Adam than on either the serpent or Eve. In addition to the explicit promise of physical death (Gen. 3:19), the ground is cursed for Adam's sake. Specifically, God promises "thorns" and "thistles" (Gen. 3:18). The Hebrew terms here, *dardar* and *qos* respectively, typically refer to thorn and thistle species found in the Holy Land.[19] Based on these terms, creationists have inferred the origin of harmful plants and plant toxins. If we relegate all animal death to a consequence of the Fall, we may reasonably assume that harmful plant products would also be a post-Fall phenomenon, especially since God gave to the animals and humans *all* the plants as food (Gen. 1:29).

THE CURSE

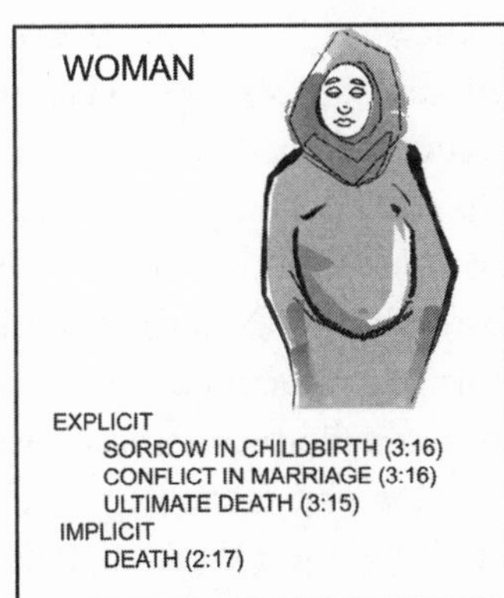

The origin of harmful plants raises an important question: Is the Curse a creation event or a modification event? In effecting the Curse, God could have created new harmful creatures, or He could merely have modified preexisting organisms. The opinion of creationists is not clear on this point, but the majority seem to favor a modification scenario, wherein a perfect creation degenerates or degrades in some way, producing the harmful organisms we have today. In chapter 10, we will return to the origin of biological imperfections, so we will not belabor the issue here.

The curse narrative culminates in Adam's personal naming of his wife (Gen. 3:20) and the first animal death (Gen. 3:21). With this first animal death, God provides His fallen children with more durable coverings than their fig-leaf aprons for their expulsion from the garden. Although we do not know the identity of the creature that died, we might wonder whether the "fig" mentioned resembled our modern fig trees. The Hebrew term translated "fig" in Genesis 3:7 is *te'ena,* used about forty times in the Old Testament. The biblical usage and related terms in Aramaic, Akkadian, and Pheonician leave little doubt as to the meaning of *te'ena*.[20] The leaves used by Adam and Eve to sew their aprons must have come from plants in the same baramin as our modern fig trees.

After their expulsion from the garden, Adam and Eve had children. Their oldest recorded child was Cain, the "tiller of the ground" (Gen. 4:2). Their next recorded child was Abel, a "keeper of sheep" (Gen. 4:2). "Sheep" is the Hebrew term for flock, *son.* In the

Pentateuch, *son* refers to flocks of goats (Gen. 27:9) and a mixed flock of goats and sheep (Gen. 30:31). Other references in the Old Testament apply *son* to flocks of sheep (1 Sam. 25:2).[21] From this broad usage, we find no reason to insist that Abel's *son* were members of our modern sheep species *Ovis aries,* but it is unlikely that Abel's *son* was composed of creatures other than members of the sheep baramin.

The curiosity of keeping sheep has been little discussed in theological literature. As we will see in the next section, God allowed people to eat animals and placed the fear of man into animals after the Flood (Gen. 9:2–3). By implication, humans ate only plants, and animals did not fear humans before the Flood. Why then did Abel keep the *son*? The new human family did not eat the animals, and their leather clothing would have been easy to obtain since animals did not fear them. Another possibility is that the *son* were kept for sacrifice.

Some scholars interpret the Cain and Abel story as teaching the importance of offering sacrifices with a pure heart, but others emphasize the unacceptability of vegetables in a sin offering. In support of the second view, scholars cite Hebrews 9:22, "Without shedding of blood is no remission." Critics rightly claim that Hebrews 9:22 refers specifically to the Levitical system, and counter with Hebrews 11:4, "By faith Abel offered unto God a more excellent sacrifice than Cain." Nevertheless, Abel's occupation as a keeper of the *son* supports a pre-Levitical sacrificial code. Furthermore, Noah's recognition of clean and unclean animals (Gen. 7:2) also supports this early sacrificial code. Consequently, both theological views are likely to be true. Abel offered the required sacrifice out of faith, but Cain offered an unacceptable sacrifice because of his lack of faith.

Strangely enough, Cain's expulsion after Abel's murder has formed the basis of a common criticism of the Genesis account. When God pronounced his punishment, Cain protested, "Every one that findeth me shall slay me" (Gen. 4:14). When Cain arrives in the land of Nod, he and his wife have a son named Enoch (Gen. 4:17). Cain's fear of meeting others who might kill him, coupled with the existence of his wife, has led some to claim that the family of Adam and Eve were not the only people on the earth at the time. Alternatively, we need merely assume that human genes were not contaminated with the modern mutational load, allowing Cain to marry his sister with no fear of deformed children. As for the people Cain feared to meet, Adam lived a total of 930 years and "begat sons and daughters" (Gen. 5:4–5). With nearly a millennium available to bear children, surely Cain would have encountered other descendants of Adam and Eve during his (extensive?) lifetime.

3.4. Flood

For young earth creationists, the Flood takes on an acute importance as the event most responsible for the physical state of our modern world. Peter tells us that "the world that then was, being overflowed with water, perished" (2 Pet. 3:6), leaving Noah's descendants with a completely different world than the one Noah and his sons had known. Most creationist biologists consider only the biological consequences of the Flood (such as genetic bottlenecks or biogeographical distribution), but we should also view the Flood as a biological event unto itself. The fossil record shows us that aquatic and marine organisms died in huge numbers during the Flood, with many baramins disappearing forever. Despite this record of death, many individuals lived through the catastrophe. The story of their survival is an integral part of the creation model.

Humans, birds, land creatures, and food plants all weathered the storm on an ark of divine design. All other organisms must have survived in the water. Pre-Flood forests became huge debris mats that floated in the oceans for decades after the Flood.[22] Plants that were not preserved for food on the ark colonized the post-Flood world from seeds or vegetative material from these mats. Microorganisms almost certainly thrived during the Flood. With so much heat and nutrients entering the water column, microbial blooms occurred frequently.

The Flood narrative begins with the gross wickedness of the antediluvian world. Seeing this evil, God chose to destroy the entire world with a Flood of water. Because "Noah found grace in the eyes of the Lord" (Gen. 6:8), Noah and his immediate family were chosen to preserve the human race, together with terrestrial animals and birds, on the ark. God commanded Noah to take two of every "kind" of cattle, creeping thing, and birds on the ark (Gen. 6:20), along with seven (presumably pairs) of clean animals and birds (Gen. 7:2–3). The words used here probably indicate land animals, including amphibians and insects. Because of the gargantuan task involved in gathering the animals, we may infer divine assistance in bringing the animals to the ark. The number of creatures on the ark forms a pivotal question in creation biology, because the number of animals on the ark will greatly influence our understanding of the **diversification** and dispersal of animals after the Flood.

The birds and animals boarded the ark according to their *min* (Gen. 6:20; 7:14), but after the Flood they left according to their *mispaha,* a Hebrew word meaning "family" or "clan" (Gen. 8:19).[23] Most commentators do not assign great significance to *min* and likewise view the *mispaha* substitution with little attention. In chapter 1, we learned that *min* might refer to divisions but should not be understood to be an invariant biological class. The association of *min* and *mispaha* here may indicate that the pairs on the ark did not always represent baramins but occasionally could have included distinct intrabaraminic groups. That is, God might have subdivided some baramins into familial lineages for survival on the ark. If correct, minor adjustments in some creationists' interpretation of the history of baramins may be necessary. Some unclean terrestrial baramins may have survived the Flood as more than a single pair.

Although this book is not a defense of the ark, the possibility of more than two members of each terrestrial baramin on the ark raises questions about the ark's capacity. Reinterpreting our understanding of baramins on the ark should not significantly alter the apologetic value of baramins for the ark's carrying capacity. Biologist Arthur Jones prefers to identify the baramin with the family level of classification, thus proposing no more than two thousand animals on the ark.[24] Whitcomb and Morris used a more conservative estimate of the size of a baramin to estimate thirty-five thousand animals on the ark.[25] If the use of *mispaha* indicates occasionally more than two animals per baramin on the ark, the number of animals was probably between two thousand and thirty-five thousand. Whatever the actual count, the ark would have sufficient space for the passengers, animals, and necessary provisions if the number of animals did not substantially exceed thirty-five thousand.

We hear often of the "two of every kind" of animal, but rarely of the crop food and seed also taken on the ark. Noah and his family certainly took a good supply of seed on the

Figure 3.3. *At 450 feet long, 45 feet high, and 75 feet wide, the scale of Noah's ark is truly colossal. The Administration Building of Bryan College, built to the same dimensions as the ark, gives us a modern-day sense of the size of this boat.*

ark not only for the duration of the Flood but for initiating agriculture after the catastrophe (e.g. Noah's vineyard in Gen. 9:20–23). God's planting of the Garden of Eden shows that plant domestication dates back to Creation itself. Since we are not told what types of plants they cultivated, we should not rule out the possibility that some crops were domesticated after Creation.

For example, Noah probably had a close relative of modern wheat or barley (or both) on the ark, because of the early appearance of these crops in the Ancient Near East.[26] In contrast, maize appears to be a very close relative of the Mexican plant teosinte and probably emerged as a separate species after the Flood. Because of the complicated interplay between the scriptural record and historical evidence, we must exercise great care when interpreting the record of plant domestication.

The Bible clearly states that all terrestrial animals perished in the Flood, but the fate of water-dwelling creatures must be inferred from extrabiblical evidence. The fossil record stands witness to the difficulty for even water-dwellers to survive the Flood outside of the ark. During the massive geological upheaval that accompanied this catastrophe, many marine and aquatic organisms lost their lives. Billions of fossil nautiloids in a single layer of the Grand Canyon illustrate the devastation wrought on the water-dwellers.[27] Many entire baramins were driven to extinction or very close to it. Most of the marine reptiles (mosasaurs, plesiosaurs, and ichthyosaurs) did not survive, along with many invertebrate groups (e.g., trilobites).

In light of the devastation of the Flood, it is remarkable that any animals survived outside of the ark, but freshwater fish provide a minor challenge to our understanding of the Flood. Freshwater fish could have survived in any number of ways. Because of the physics of water mixing, large pockets of low-salinity water could have persisted throughout the duration of the Flood. Alternatively, modern aquatic fish could have descended from marine ancestors that colonized bodies of fresh water after the Flood. More fish baraminology research is necessary before definitive answers can be given.

Despite the devastation seen in the fossil record, the Flood did not destroy all life in the water. Some organisms not only survived but even thrived. The pre-Flood forests were ripped up by the Flood, producing giant, floating debris mats. The material deposited from the debris mats became our present coal fields.[28] Based on the amount of coal in the modern world, the size of the debris mats must have been enormous. These large mats undoubtedly provided an excellent haven for much plant material to be kept floating without direct contact with the ocean. Other organisms almost certainly thrived during the Flood. The eruption of the "fountains of the deep" and submarine volcanoes infused the oceans with hot, microbe-rich material. For example, archaebacteria generated chemicals which can still be found in layers of shale.[29]

3.5. Post-Flood Recovery

The Bible reveals little information regarding the recovery of the earth following the Flood. We have a few enigmatic passages that give us tantalizing clues about the conditions immediately after the Flood. Besides the linguistic confusion at the tower of Babel, the best-known post-Flood phenomenon is the rapid drop in the life spans of the patriarchs recorded in Genesis 11:10–32. Other oddities include passing references to weather conditions. Lot chose to move to Sodom because the plain of the Jordan River was "well watered" (Gen. 13:10), but today, the same region is a barren desert.

The Book of Job describes events that took place during the patriarchal period, possibly contemporaneous with Abraham. Creationists are most familiar with Job for the lengthy descriptions of Behemoth and Leviathan in Job 40–41. Behemoth is the most identifiable of the two, nicely fitting the description of a large dinosaur. Job also appears to be fixated on chilly weather. In the entire Old Testament, Job contains 19 percent of the references to *snow,* 67 percent of the references to *ice,* 33 percent of the references to *frost,* 20 percent of the references to *cold,* and the only reference to *frozen.* To make sense of these biblical tidbits, we turn to a variety of scientific models developed in the past twenty years.

The story of the post-Flood recovery process began with the cooling of the hot oceans. During the Flood, large amounts of lava reformed the ocean floor as the original one fell into the mantle.[30] By the end of the Flood, hot rocks on the bottom of the ocean increased the overall temperature of the ocean by 20°C, producing a chain reaction of climatic events.[31] First, atmospheric models show that a giant hurricane probably formed over the north Atlantic as ocean waters evaporated into the atmosphere at an unprecedented rate. Because the warm ocean waters continually fed this hurricane, it did not dissipate as our modern hurricanes do. Instead, this hurricane raged for decades as the oceans gradually cooled.[32]

The warm ocean waters also increased rainfall worldwide. In lower latitudes, high precipitation led to the blooming of areas that are now arid deserts. When Lot saw the well-watered plain of the Jordan, he was looking at the effects of the post-Flood climate.[33] Similarly, evidence of Sahara grasslands[34] and a water-eroded Sphinx[35] fit nicely into this post-Flood period.

In higher latitudes, the heavy precipitation produced a different result. Because of the c and volcanics occurring during and after the Flood, the post-Flood atmosphere was probably often filled with a thick cloud of ash and soot. As a result, the continental interiors received less solar heating than they do now.[36] The evaporation of ocean water gradually

Figure 3.4. *Hurricane Elena over the Gulf of Mexico, photographed in September 1985 from the space shuttle. A giant hurricane like this one persisted for decades after the Flood in the north Atlantic.*

cooled the oceans. As this cooling of the land and water continued, precipitation fell as snow in the higher altitudes and at certain continental positions. Eventually this snow buildup produced a single ice advance (known to non-creationists as various ice ages).[37] When Job spoke of the snow and the cold, he spoke as an eyewitness to the time of the ice buildup and advance.

The high precipitation rates also generated giant inland lakes. Rapid glacial melt added to the lake development. These lakes often swelled, overflowed their banks, and carved tremendous canyons. For example, three huge lakes covered vast areas of four states in the United States: Utah, Arizona, New Mexico, and Colorado. The breaching of these lakes through the Colorado plateau carved the Grand Canyon.[38] With the melting of continental glaciers, the sea level rose. Water from the Mediterranean inundated the Black Sea, flooding the shoreline settlements.[39] The same rise in the oceans flooded the Bering land bridge, severing the land connection between Asia and North America. In the eastern United States, meltwater from the ice helped to carve the thousands of caves in Tennessee and Kentucky.

During this time, regional geological catastrophes struck the earth frequently. India collided with Asia and formed the Himalayas after the Flood. Giant volcanoes erupted throughout this period. In North America, a substantial portion of Yellowstone National Park sits on the collapsed crater of a single, gargantuan volcano.[40] Volcanoes in the oceans produced thousands of islands all over the world. Other land masses, such as the Isthmus of Panama, rose more gently out of the oceans. Prior to the rising of Panama, North and South America were not directly connected.[41]

The combination of the ice surge, the cooling oceans, and the settling of debris from the atmosphere catastrophically altered the global wind patterns. Immediately after the Flood, the warm oceans produced a specialized wind pattern significantly different from the modern one. As the oceans cooled, there came a critical point, a few centuries after the Flood, when the post-Flood atmospheric cycle rapidly assumed the cycle we have today. In perhaps less than a decade, the world's climate changed. Areas approximately 20° north and south latitude dried out and became deserts. At the same time, regions at 20–50° latitude became temperate. Simultaneously, the cooling of the oceans generated a global reduction in precipitation, continuing the overall drying trend.[42]

These global geologic and atmospheric changes had tremendous biological repercussions. Debris mats and log jams floating in the oceans provided organisms with a means of rapid dispersal along ocean currents.[43] Commensurate with the changing environment, the fossil record reveals that baramins themselves changed, often in fits and starts, swiftly producing a high diversity in many baramins.[44] In the immediate post-Flood period, baramins showed abilities to adapt and disperse, but these abilities waned as time proceeded.[45] By the time of the ice surge, organisms no longer adapted to the changing conditions. Despite surviving numerous localized geological catastrophes, species everywhere died as the global wind patterns shifted for the last time. We will refer to these extinctions as Decimations.

Early dispersal from the ark was accomplished by rafting on debris mats and logjams as they floated along ocean currents. The size of the mats provided a certain level of stability that allowed normally terrestrial animals to live on them with little discomfort. Since they were composed of plant material, the mats were the centers of plant recovery, drawing

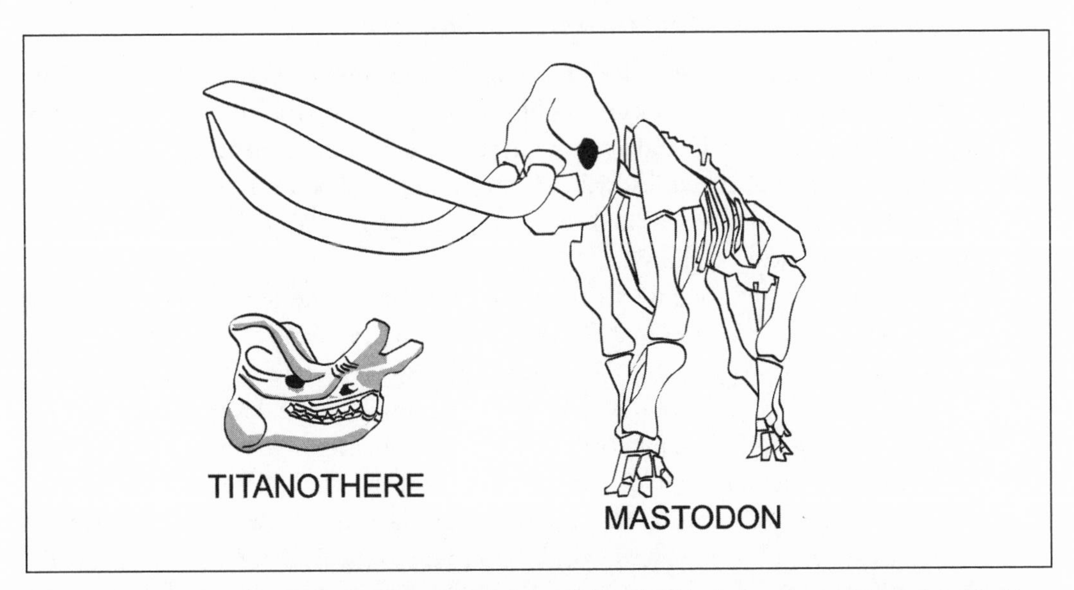

Figure 3.5. Post-Flood diversity was much higher than it is today. North America was once inhabited by titanotheres and mastodons, as well other species of elephants, rhinoceroses, camels, lions, and horses. Today, these groups are either extinct or naturally found only on other continents.

herbivores (and thus carnivores) to them. As time progressed, many of the logs that formed these mats gradually became waterlogged and sank. The destabilization of the mats due to the sinking of constituent logs prevented further dispersal.[46]

Because of the frequency of regional catastrophes after the Flood, the post-Flood fossil record preserves the history of biological recovery remarkably well. In many mammal baramins, actual fossil series can be found, showing a transformation from small ancestors to their larger and more diverse descendants. Small animals would be a great choice for preservation on the ark because they needed less room than larger members of the same baramin and they thrived on the warm, post-Flood earth. The best-known example of the post-Flood diversification is found in the fossil horses,[47] but camels, elephants, rhinoceroses, cats, rodents, and just about every other mammal group show similar fossil trends.

The fossils of North America beautifully demonstrate the effects of the Decimations. Soon after the Flood, as the continent was invaded repeatedly by colonizers rafting in from Europe, large forests dominated the landscape in the southern latitudes. Herbivores from this period existed predominantly as "browsers," nibbling on the foliage of trees and shrubs. In more northerly and inland regions, larger and hardier animals, such as mammoths, found a pleasant environment in the mild, uniform climate. Throughout this period, local catastrophes preserved the remains of many of these animals. When the atmosphere cycles changed, the middle latitudes dried out and the flora shifted from forest to grassland. Herbivores that grazed on grass flourished at this point. The subsequent ice surge and the new atmospheric cycles all conspired to doom most of the larger organisms in North America. Today, no North American mammoths, horses, or camels survive, despite thriving those pivotal centuries after the Flood.

3.6. Chapter Summary

Genesis lays out for us a very clear and unmistakable outline of earth history. While some chose to debate the obvious meaning of the text, it is important to accept the assumption that the Bible speaks clearly and authoritatively on all matters it mentions. With that in mind, however, the careful creationist will review a number of sources to make certain an accurate translation of biblical terms and contexts is maintained. We may infer four biblical events that greatly impacted organisms:

1. Creation occurred roughly six thousand years before the present day. According to Genesis 1, God completed Creation in only one week.
2. The Fall of man is arguably the most significant event in the history of life on this planet. Many attributes of modern living things testify to the devastation of the Fall. The Curse stems from human disobedience. Creationists use the effects of the Curse to explain a multitude of biological phenomena such as harmful plants and animals.
3. The Flood takes on an acute importance as the event most responsible for the physical state of our modern world. According to fossil records, it was an important biological event because entire baramins were destroyed forever.
4. Post-Flood recovery remains somewhat of a mystery to modern creationists. We have a few enigmatic passages that give us clues about the conditions immediately

after the Flood. Early dispersal from the ark was accomplished by rafting on debris mats and logjams as they floated along ocean currents. Of course, the number of species living on the ark provided the basis for the post-Flood recovery of land animals and birds.

Review Questions

1. Where do we find the outline of earth history in the Bible? Briefly describe that history by using biblical references.
2. According to the chronology used in this book, roughly how old is the earth?
3. List the explicit curses in the Curse.
4. List some possibilities for the identity of the serpent in Genesis 3.
5. What is the origin of harmful plants and animals?
6. What does the story of Cain and Abel tell us about the sacrificial system?
7. Was the Flood entirely destructive to life?
8. How do crops and seeds change our understanding of post-Flood recovery?
9. How do we know that crop domestication dates to Creation?
10. Define decimation.
11. Provide English translations to the following Hebrew terms:
    a. *tannin*
    b. *behema*
    c. *eseb*
    d. *rms*
    e. *hayya*
    f. *nahas*
    g. *te'ena*
    h. *remes*
    i. *ets*
    j. *mispaha*

For Further Discussion

1. How does the age of creation change our understanding of the pattern of life?
2. Should we invoke miraculous events to explain the recovery of the earth? Why or why not?
3. What role does the Fall play in our understanding of baraminology?
4. Does the length of a "day" in Genesis 1 have any pertinence to our understanding of baraminology?
5. Based on information provided in Genesis, what can we conclude about the pre-Flood climate? What impact would climate have on our understanding of baraminology?
6. Why is the issue of carnivory difficult to incorporate into an understanding of the pattern of life?

7. Of all the creatures God created, why would Satan choose the serpent as the means for tempting Adam and Eve? Can you make any inferences about the pattern of life based on that choice?
8. How did the earth begin recovering from the effects of the Flood while Noah and company were still on the ark?
9. Considering the changes that occurred during the post-Flood period, speculate on how organisms might have reacted to the major climatic changes.
10. What does the recovery of an olive leaf by Noah's dove mean to baraminology?

SECTION II

Methodology

CHAPTER 4

Gathering and Interpreting Biblical Data

4.1. Introduction

This chapter opens a section on the methods of baraminology. In the following chapters, we will present scientific and statistical methods for identifying modern members of baramins. This chapter focuses exclusively on gathering data from the most reliable source we have, the Bible.[1] Although the Word of God is not a science textbook, it speaks authoritatively and without error. By its own claim, the Bible is our primary source for creationism. "Through faith we understand that the worlds were framed by the word of God" (Heb. 11:3), and "faith cometh by hearing, and hearing by the word of God" (Rom. 10:17). Biblical creationism flows from suppositions held in faith, and faith comes by the Word of God. Because of the value of Scripture, all baraminology studies ought to be framed within its revealed truth.

Many people devote their lives to studying the Bible and writing about their studies in countless books, articles, and sermons. Like formal training in any discipline, professional study of the Bible brings insight and experience that can be gained in no other way. Ideally, baraminologists should draw upon that valuable experience by seeking the assistance of biblical scholars. Since collaboration between scientists and Bible scholars is unfortunately rare, baraminologists could also proceed with the limited resources available to them, recognizing that their own conclusions are only preliminary.

This chapter will present information and techniques for the amateur Bible student who wishes to evaluate biblical references that relate to baraminology. As we use these techniques, we must remember that we amateurs can (and probably will) make mistakes. If you cannot find a professional Bible scholar to be your collaborator, try to find a scholar who might at least review your conclusions to point out errors or inconsistencies prior to publication. In this way, we can minimize the publication of error.

Biblical studies relevant to baraminology usually focus on the identity of animals or plants in the Bible. Baraminologists look for references to organisms in the Scripture, focusing on the first appearance of the creatures and their relation to the term *min* (if any). Important information can be gleaned from the original Hebrew texts, from ancient and modern translations, and from scholarly commentaries. Combined with the historical considerations outlined in chapter 3, a rigorous study of the biblical text greatly enlightens baraminology.

4.2. How the Bible Aids Baraminology

Baraminologists look to the Bible for assistance in identifying baramins and understanding their history. As we discussed in chapter 2, the separate days of creation imply that fundamental discontinuities exist between groups of organisms. Minimally, the flying creatures, swimming creatures, and land creatures form separate, discontinuous groups. This type of evidence can aid in the identification of baraminic groups, such as apobaramins. Additionally, early references to organisms amplify our knowledge of the history of baramins. Because so many questions about baramins and their ancestry remain unanswered, we need strong biblical studies to expand our understanding of baraminic history.

In chapter 3, we also noted that some of the terms in Genesis 1 could be descriptive rather than taxonomic. Thus, it is possible that woody and herbaceous plants could occur in the same baramin. In other cases, terms such as "flying creatures" do seem to define apobaraminic classes. Biologically, powered flight does seem to exist in strong discontinuity to the gliding or creeping things; however, if "herb" and "tree" describe individuals, it is possible that "flying things" could also describe individuals. Furthermore, members of a baramin might have lost the ability to fly during their history, but in such cases, vestiges of the ability to fly would probably remain. Consequently, we should not *a priori* reject the possibility of a baramin in which some members can fly and others cannot. Such a baramin certainly exists among modern insects, but the existence at the time of creation of a baramin in which some members could fly and some could not cannot be determined conclusively.

Our difficulty interpreting the terms of Genesis 1 should remind us of the emphasis on holistic evidence in the refined baramin concept. Baraminic groups should not be defined on the basis of a single characteristic, such as wood or flight. Doing so invariably leads to difficulties in reconciling biology with our narrow view of Scripture. By maintaining an open and critical mind, we can avoid the pitfalls of reductionistic biology. Careful attention to both biology and the Scripture should help us to discriminate between terms used as classification and terms used as description.

Studying the history of baramins poses fewer difficulties than examining the record of creation. In such cases, the baraminologist begins with scientific evidence for the existence of a baramin and proceeds to the Scripture to find evidence of the history of the baramin. Such evidence usually takes the form of references in Genesis or Job (which was written during or before the time of Abraham). Records of baramin members early in the post-Flood period allow us to gauge the rate of diversification in a terrestrial animal group. We can also find evidence of the post-Flood recovery of a non-terrestrial or non-animal group.

For example, the first mention of a camel in the Bible occurs in Genesis 12:16: "And he entreated Abram well for her sake: and he had sheep, and oxen, and he asses, and menservants, and maidservants, and she asses, and camels." This event took place no more than 367 years after the Flood. Job also possessed 3,000 camels before his suffering and 6,000 after (Job 1:3; 42:12). Although Job cannot be dated precisely, he lived approximately at the same time as Abram or even earlier.

Since the Hebrew word for camel, *gamal,* occurs fairly frequently in the Old Testament (fifty-four times), we may safely assume that *gamal* refers to a modern camel, and by reference to the present distribution of camels, to the dromedary camel. Since both Job and Genesis refer to the *gamal* as a domesticated animal, the domesticated dromedary probably existed shortly after the Flood. The fossil record of the camel baramin shows a rich diversity similar to that of the horse. The camels present on the ark were probably small and very unlike modern dromedaries. Consequently, Job, Abram, and the pharaoh of Egypt may have been among the first people to domesticate and use the modern camel species.[2]

A greater diversity of references to grasses appear in the biblical record. Job mentions both wheat and barley (Job 31:40), as well as numerous agricultural terms, such as "grain" (Job 5:26 NKJV), "fodder" (Job 24:6 NKJV), and "heads of grain" (Job 24:24 NKJV). Other terminology in Job shows evidence of linguistic evolution from specific agricultural words to a more general usage. In Job 39:4, the term *bar* technically means "grain," but God seems to use it in reference to the "open field," as the NASB translates it.

Like the camel references, early grass references place constraints on the timing of grass diversification. Several distinguishing features, however, lead us to a different interpretation of the grass baramin's history. First, the camel baramin has many fewer species than the grass baramin (~200 vs. ~10,000). Second, as an unclean terrestrial animal, camels were probably represented by two individuals on the ark, but crop plants (such as the grasses) and their seed were probably carried in abundance on the ark. We may explain the large diversity of the grass baramin more easily if we assume that many of the major grass groups had diversified by the time of the Flood and were preserved separately. As a result, the sophisticated agriculture of Job's time probably represents both pre-Flood diversity and post-Flood crop domestication.[3]

As illustrated in the examples of the camels and the grasses, the Bible can be of immense help in understanding the history of baramins, even when it does not illuminate baraminic boundaries. When examined carefully, the information in the Bible might reveal information that can be found no other way. As we work on baraminology research, we should keep in mind a list of relevant questions for our biblical analysis:

1. Is my group mentioned in the Bible? Remember large groups (like "creeping things") or indirect evidence (by-products or diseases).
2. When did my group first appear in the Bible? Since most baraminic diversification ceased shortly after the Flood, references in Job, Genesis, or Exodus reveal the most information about diversification rates.
3. How did my group survive the Flood? Since the Flood represents an enormous population bottleneck, we should try to determine the number of ancestors our group has. Terrestrial animal baramins may differ markedly from marine animal or plant baramins because of the difference in the number of survivors.

4. What does the first scriptural appearance and the Flood survival tell me about my baramin? As the camels and grasses illustrate, the answer to this question varies from group to group.
5. Does my baramin show signs of "groaning and travailing" under the Curse? We mention this question now because it relates to our study of the Scripture, but we will return to it for a fuller discussion in chapter 11.

Because these questions focus so closely on the early post-Flood history of baramins, we will primarily discuss Old Testament resources, while briefly mentioning a few resources for the New Testament.

4.3. Interpreting the Bible

We practice interpretation every day. Whether we watch television, read the newspaper, or browse the internet, we encounter various forms of communication from which we infer meaning. Most of the time, we interpret meaning without any thought to the process of interpretation itself. Scholars call the process of interpretation **exegesis** (from a Greek word meaning "to draw out"). Exegesis aims to draw out the meaning of the text, to discern meaning in the words that the author wrote. In gathering biblical evidence to answer the questions of the previous section, baraminologists must engage in exegesis. Here, we will review some very basic considerations and principles that guide good exegesis.

We must begin our discussion with a very simple question: What do we mean by the "meaning" of a text? We often refer to "the" meaning of a text, as if it had only one. Historically, this has rarely been the position of Bible interpreters. Augustine believed that the primary meaning of a text resides in its allegorical or figurative meaning. In the Middle Ages, theologians sought multiple meanings of texts, from the plain meaning to the heavenly symbolism. Medieval Jewish scholars, concerned with obedience to the law, tended more toward straightforward interpretation of the Bible. Martin Luther advocated the straightforward meaning of the words themselves as the only valid meaning of the text. The twentieth century brought us post-modernism and deconstructionism, which deny the existence of objective meaning inherent in the text.[4]

In this confusion of meaning, the Bible itself reveals multiple valid meanings for its own writings. As the New Testament writers interpreted the Old Testament, they frequently departed from what we understand to be the straightforward meaning of the text. For example, when Paul discussed the righteousness attained by the Gentiles through faith, he quoted Hosea 2:23, "I will have mercy upon her that had not obtained mercy; and I will say to them which were not my people, Thou art my people"; however, Hosea spoke of Israel when he originally wrote those words. Paul found deeper meaning in the original passage, and since both passages are Scripture, both meanings must be correct.

To detect the various layers and facets of meaning contained within a text, we can approach the problem from a number of perspectives. We could try to seek the wording of the original autograph of Scripture by comparing manuscripts. We could evaluate the words, grammar, and language used in the passage. We could consider the style or genre of the larger work from which the passage is taken. We could examine the historical, geographical, or cultural context in which the text was written. We could even study the writer himself: Where did the writer obtain the information he wrote, and what themes did he intend to convey with his writing? Each of these areas present relevant data for the interpretation of a biblical passage.

Baraminologists attempt to identify references to organisms and interpret them primarily in their historical context. This involves studying the meaning of Hebrew (and occasionally Greek) words, understanding their usage in various passages, understanding the context of the passages, and placing the passages in their correct historical setting. In the remainder of this chapter, we will review basic resources and techniques for gathering data about biblical texts, followed by a brief discussion of how to draw conclusions from the data gathered. Once again we must emphasize that training and experience yield ability in biblical exegesis that novices and amateurs do not have. Consequently, we amateur Bible students ought to seek assistance and advice from professionals who are willing to help.

4.4. Finding Organisms in the Bible

We begin our biblical study by locating references to our group of interest in the Bible. To find terms for organisms, we must put our biological classification into ordinary terms that people are most likely to use in their everyday conversations. These words can include references to the organisms or to by-products of the organisms (such as cloth made from plants or diseases caused by microbes). For example, several Hebrew words apply to "sheep," including *taleh* (suckling lamb), *kebes/keseb* (lamb or sheep), *rahel* (ewe lamb), *asarot* (lambs), and *ayil* (ram).[5] Since the Bible is not a science text, all of these words occur regularly, but no simple summary of words for sheep appears.

Similarly, the Hebrew terms used for grasses include a variety of words that refer to different species as well as different growth habits (like the terms for sheep). The grass family Poaceae includes many important crop plants, such as rice, maize, wheat, barley, and rye. Some ancient Hebrew words unambiguously refer to species or genera of grasses, such as *kussemet* ("rie," Exod. 9:32), *hitta* ("wheat," Gen. 30:14), and *se'ora* ("barley," Job 31:40). Other words arise from the obvious agricultural importance of the grasses, such as *sibbolet* ("ears of corn," Job 24:24), *dagan* (KJV: "corn," NIV: "sheaves," Job 5:26), and *bar* (KJV: "corn," NASB: "open field," Job 39:4).[6] In studying the grasses in the Bible, one might also consider the by-products of grasses, such as *flour* (Exod. 29:2) or *bread* (Gen. 14:18), or words referring to agricultural activities, such as *harvest* (Exod. 34:22) or *glean* (Ruth 2:2). All of these words bear upon our understanding of grasses in the Bible.

With a preliminary list of terms in hand, the search for words in the scriptural text can begin. Most frequently, amateur biblical scholars begin with one of two resources: *Strong's Exhaustive Concordance of the Bible* (available in a new 1997 edition from Thomas Nelson Publishers) and *Young's Analytical Concordance of the Bible* (available in a 1993 edition from Hendrickson Publishers). Both *Strong's* and *Young's* contain a concordance of references to all the English words in the King James Version of the Bible (a version of *Young's* that refers to "Young's Literal Translation of the Holy Bible" also exists).

Strong's Exhaustive Concordance of the Bible also provides short Hebrew and Greek dictionaries, with the Hebrew and Greek words correlated by number to the English words in the concordance. In *Strong's,* each English word lists both the verses in which it occurs and the Strong's number which relates it to the corresponding entry in the Hebrew or Greek dictionary. Strong's numbers are often used in other biblical reference material. Words added to the English text for clarity by translators are not included in the *Strong's* dictionaries.

Young's Analytical Concordance resembles *Strong's* in many respects but also offers several features that *Strong's* lacks. Instead of a numbering system and Hebrew dictionary, *Young's* groups the references to each English word by the original word that it translates. Thus, under any Old Testament word, the Hebrew words and their usual meanings are listed, with the references in which the words occur listed under each Hebrew word. In this way, *Young's* is slightly easier to use than *Strong's* since the additional time to look up the various meanings of the Hebrew terms is unnecessary with *Young's. Young's* also offers a brief Hebrew lexicon, in which the Hebrew words are listed together with the words used to translate them in the English Bible. In this way, *Young's* can also be used as a kind of concordance to the original Hebrew Bible. We will discuss the importance of this feature in the next section.

For rapidly locating references to words without the Hebrew or Greek cross-referencing, nothing surpasses the convenience and speed of the Bible Gateway (http://www.biblegateway.com). Other electronic Bible products can be purchased, but good products of this type can cost hundreds of dollars, while the more affordable versions usually include only a few translations. In contrast, the Bible Gateway contains eleven English translations with free searching to anyone with a connection to the internet. Significant English translations available at the Bible Gateway include the King James, New King James, New International, New American Standard, and Revised Standard. Additionally, the Bible Gateway also provides searching in Bibles translated into seventeen other languages, including the Latin Vulgate and Luther's 1545 German translation.

The Bible Gateway provides two types of searches: reference and word searches. Reference searches can display a verse in any language (e.g., Gen. 1:1). Word searching resembles a concordance, but the user can search for words in any translation (English or otherwise). Partial word searches can also be performed, and the searches can be limited to specific books. With both types of searches, the results give the reference and text for each occurrence, with the search term highlighted in bold text. The text viewing options (accessed by clicking the appropriate reference) allow the user to view the verse, adjacent verses, or the entire chapter for each reference. The text view presents links that allow the user to "toggle" back and forth among the different translations or to display more than one translation at the same time.

These three sources provide a powerful means of discovering references in the Hebrew Bible. Scholars refer to the numbering system of *Strong's* in many other resources, and *Strong's* dictionary provides a very useful guide to the general meaning of Hebrew terms. Using *Young's* as a Hebrew concordance proves valuable in locating references to organisms, especially when the list of organismal terms is incomplete. Finally, although providing no Hebrew or Greek translation, using the Bible Gateway can significantly accelerate discovery of biblical references and comparison between translations.

4.5. Verifying the Translation

Locating terms that refer to organisms of interest in an English translation of the Bible does not necessarily mean that the Bible actually refers to those organisms.[7] Conservative Bible scholars universally agree that the only truly inspired copies of Scripture are the original autographs, the documents produced by the original authors. We possess none of these documents today, so it is possible that grammatical or spelling errors have crept into the text due to copying mistakes. These errors do not threaten the doctrine of inerrancy, but they can muddle the meaning of passages where they occur. If one manuscript contains a reference to an organism that another manuscript omits, we must decide which manuscript is more faithful to the original autograph. All biblical scholars should strive to come as close to the original autograph as possible.

The primary Hebrew text used in translating the Old Testament is the Masoretic text. Standardization of the Hebrew text of the Old Testament began in the second century, partially as a reaction against the Christian adoption of the Septuagint. By the sixth century, a group of Jewish scholars called Masoretes introduced a careful method of counting words and characters to insure transcriptional fidelity. The Masoretic text that translators use today is a further standardization of various Masoretic texts dating from approximately 1000.

Another problem that we will encounter in biblical study is translation difficulty. Deciphering biological terms often poses substantial difficulties to Bible translators. When a Hebrew word appears frequently in the scriptural text, translators infer meaning from its usage and context. If a term occurs only once in the Hebrew text of the Bible, the meaning can only be guessed by the context of the single usage or by proposing relationship to other terms. Because of these types of translation difficulties, we devote this section to strategies that amateur Bible students can use to verify the translation and investigate translation difficulties. We will progress from the easiest strategies to the more complex.

As a simple first approximation to detecting translation difficulties, we can compare various English translations. If each version has the same word for the Hebrew term of interest, translation disagreements probably do not exist. To do this correctly, we must first identify literal English translations. Some versions are very faithful to render the Hebrew or Greek text literally into English. Other versions use a more thought-for-thought style of translation to make the English Bible more readable. Still other versions are paraphrases and not translations at all. Since our goal is to understand translation differences, we must compare translations that are close to the original, literal text rather than the more free-style conceptual translations.

The King James Version (KJV), originally published in 1611, is by far the most popular English Bible in all of history. Authorized and actively coordinated by King James I of England, the translation team that produced the KJV consulted previous English versions, as well as important Hebrew and Greek texts. The resulting translation far exceeded previous English translations in readability and accuracy. Subsequently, the KJV has undergone several revisions throughout its history in order to update the rapidly evolving idioms of the English language. In all editions (including the recent New King James Version), the KJV tends to render the original languages literally.

Publication of English Bible translations has increased exponentially in the past two centuries. This growth stems in part from discovery of earlier manuscripts and in part from increased study of Hebrew, Greek, and related languages. Newly-discovered ancient manuscripts primarily impact the translation of the New Testament, but the discovery of the Dead Sea Scrolls in 1947 and 1948 has broadened our understanding of the Masoretic Hebrew text. Modern literal translations include the New American Standard (NASB), the Revised Standard (RSV), and the New Revised Standard (NRSV). Other translations, such as the New International (NIV), take more freedom in rendering the concepts, but not necessarily every word, of the original texts, but the NIV represents very recent scholarly opinion and should be consulted.

Using the Bible Gateway, we can create a chart of translation comparisons very quickly and efficiently. For example, little time is required to discover that the Hebrew term *dohan* in Ezekiel 4:9 is translated "millet" in the KJV, NKJV, NASB, RSV, and NIV. The uniformity of translation here implies that a consensus of opinion probably exists and that the translation is not controversial. In contrast, many difficult terms exist in the lists of clean and unclean animals in the Levitical law. For example, consider Leviticus 11:18. The KJV renders this verse, "And the swan, and the pelican, and the gier eagle," but the NIV reads, "The white owl, the desert owl, the osprey," while the RSV has, "The water hen, the pelican, the carrion vulture." Even the NKJV has a different translation, "The white owl, the jackdaw, and the carrion vulture." This much variation in translation alerts us to the fact that these words are poorly understood, and no English translation should be considered reliable.

Another excellent way to get closer to the original autographs involves consulting ancient translations or alternative versions of the Hebrew Masoretic text. In doing so, we must exercise care and discretion, since an older manuscript is not necessarily superior to a modern one. Alterations and strange insertions are well-known in ancient translations, but the choice of words in these Bible versions can add to the overall evidence for a particular definition or reading. Ancient translations and versions include the Septuagint, a Greek translation of the Old Testament, and the Vulgate, Jerome's Latin translation, and the Samaritan Pentateuch, a Hebrew version of the books of Moses preserved in the northern kingdom of Israel.

The origin of the Septuagint cannot be determined with certainty because of the legends that have grown up around it. According to the most common story, seventy Jewish scholars translated independently yet produced the same Greek wording in their versions. Based on historical records, we can estimate that Jews in Alexandria created the Septuagint around 200 B.C. Although the Greek translation varies between literal and almost commentary, the Septuagint provides a witness to early Hebrew texts independent of the Masoretic text. Zondervan publishes *The Septuagint and Apocrypha: Greek and English,* as compiled and translated by Lancelot Brenton, originally published in 1851. In Brenton's Septuagint, the English translation and the Greek text appear side-by-side for simplicity of use.

Latin texts began to appear in the second century, but a complete Latin translation did not appear until the fourth century. Pope Damasus I commissioned Jerome to produce a standardized Latin text for liturgy. Jerome's translation, the Vulgate, cannot be viewed as an authoritative witness to ancient Hebrew or Greek texts, because Jerome relied on the Septuagint and Latin texts to produce the Vulgate. In difficult places, Jerome relied most heavily on the Septuagint. Nevertheless, if used with appropriate caution, the Vulgate may occasionally offer insights into the translation of difficult terms. The Bible Gateway includes the complete Latin Vulgate, and like all of their Bibles, users can rapidly locate references and toggle between translations.

When the northern ten tribes split from Judah at the time of Jeroboam and Rehoboam, scribes in the north preserved copies of the Hebrew scriptures independently from the southern texts, which became the well-known Masoretic text. Since most of the historical, poetic, and prophetic books were written after the split between Judah and Israel, Samaritan scribes preserved only the Pentateuch. The Samaritan Pentateuch contains approximately six thousand variations when compared to the Masoretic text, mostly spelling differences. The Samaritan Pentateuch is not readily available to the amateur Bible student, but we ought to know and understand the history of it because we will encounter it when we dig deeper into other types of references, such as Bible dictionaries and commentaries.

In addition to comparing translations, the amateur Bible student may also consult Hebrew dictionaries, theological word books, and other scholarly linguistic resources. Two multi-authored Hebrew dictionaries furnish biblical researchers with excellent and comprehensive summaries of the primary terms found in the Hebrew Bible. Eerdmans publishes the *Theological Dictionary of the Old Testament* (TDOT), edited by Botterweck, Ringgren, and Fabry. Currently at eleven volumes and still not complete, TDOT is acknowledged by biblical scholars as one of the most comprehensive resources on Old Testament Hebrew. Because of the prohibitively high price of TDOT, it may be best to locate a copy in a nearby university or seminary library.

As an alternative to TDOT, Zondervan publishes the *New International Dictionary of Old Testament Theology and Exegesis* (NIDOTTE), edited by VanGemeren. Written by a host of conservative Hebrew scholars, the five-volume NIDOTTE includes an index of Strong's numbers, allowing correlation between the concordance and dictionary entries. Unlike the TDOT, NIDOTTE is a complete dictionary of Hebrew terms and is more affordable for the amateur Bible student.

Both of these dictionaries present essays on the major words found in the Hebrew Old Testament. The authors review the linguistic usage in the Old Testament, related Hebrew terms, and similar words in other ancient Near Eastern languages. The essays also include bibliographies for further research and study. Dictionaries also offer the amateur student the ability to obtain information that would otherwise be inaccessible. For example, dictionary authors often mention variant wordings found in the Samaritan Pentateuch or in manuscripts of the Dead Sea Scrolls, resources difficult for the amateur to access and use properly.

Of lesser extent and scope than the dictionaries are theological wordbooks, such as the two-volume *Theological Wordbook of the Old Testament,* edited by Archer, Waltke, and Harris and published by Moody Press. With articles similar in style to TDOT and NIDOTTE, the *Theological Wordbook* covers far fewer words than either of the dictionaries. The chances of finding more obscure terms in the *Theological Wordbook* are lower than in the dictionaries, but words that are present are covered with similar depth.

The amateur student seeking even more depth regarding Hebrew terms should turn to theological and biblical studies journals. Many of these journals have a long history of publication and contain hundreds of articles on the interpretation of words in the Hebrew Old Testament. We should expect at least a few articles on the meaning of biological terms in this large body of literature. Furthermore, the professional literature can often be a window into the kinds of disagreements that exist over the translation of particular Hebrew words. Knowledge of these disagreements helps the baraminologist to remain within the boundaries of acceptable application of biblical information. For example, if two interpretations of a single word are possible, and we as baraminologists insist on one while ignorant of the other, our baraminological interpretations may suffer.

One accessible source for literature references is the bibliography given in the dictionaries or wordbooks. If the term of interest represents a crucial part of a baraminological argument or interpretation, obtaining the references listed in the bibliography might be warranted for a fuller understanding of scholarly opinion. Once articles from the primary literature have been obtained, scan their bibliographies for further works of interest. For original literature searches, the American Theological Library Association (ATLA) provides an enormous database of professional religious and theological literature. The full ATLA database contains over a million records, but the biblical studies division of the database narrows the coverage to articles dealing directly with the Bible. The ATLA biblical studies database may be searched by keyword or by Bible reference. Many seminary libraries make ATLA available to their patrons, or the biblical studies ATLA database may be purchased at http://www.atla.com.

4.6. Commentaries

The most common and familiar resource available to amateur Bible students are Bible commentaries. Commentaries run the gamut from very general reviews of the whole Bible (e.g., *Matthew Henry's Commentary on the Whole Bible*) to very specific and detailed analyses (e.g., Westermann's three-volume commentary on Genesis). The purposes of commentaries vary from providing encouragement to fellow Christians to serving as a pastoral resource to critically examining the biblical text. The perspective of commentaries also ranges from solemn respect for biblical inerrancy to treating the Bible like any other ancient document. With discretion, the baraminologist can gain much valuable information and insight from commentaries.

Since baraminologists need more detail than the average pastoral commentary provides, we will focus our discussion on scholarly and textual commentaries. The key to wise use of commentaries is variety. Relying solely on one commentary will introduce biases of the commentary authors into the baraminological analysis. To prevent this, always consult multiple commentaries. Here, we will simply review a few commentaries on Genesis that we have found helpful.

The aforementioned commentary on Genesis by Claus Westermann furnishes the reader with an extremely detailed and richly documented commentary on the Book of Genesis. Westermann divided his commentary into sections corresponding to self-contained narratives in the biblical text, and each section begins with an extensive bibliography. Like the dictionaries and wordbooks, a good bibliography allows the amateur student to delve further into the scholarly literature. Westermann's commentary suffers from his adherence to European higher criticism, treating the text of Genesis as a conglomerate of myth and priestly editing. Nevertheless, he often has valuable insights that we have found nowhere else. The baraminologist must discriminate valuable insights of liberal commentaries from their unacceptable conclusions that arise from a lack of respect for biblical inerrancy.

In *I Studied Inscriptions from Before the Flood,* Hess and Tsumura collected a sampling of scholarly articles on the interpretation of Genesis 1–11. The articles include comparisons of the Genesis texts with other Ancient Near Eastern documents, discussions of archeological and historical studies, and literary analyses of specific passages. Taken from biblical studies journals, the articles range in date from the 1950s to the 1980s. Because of the variety of perspectives on biblical inerrancy implicit in the articles in this volume, the baraminologist should exercise discretion in studying these sources.

For an unusual perspective, consider the Jewish Publication Society's *Torah Commentary.* The JPS Genesis commentary, written by Sarna, contains the full Hebrew text of Genesis and a scholarly discussion of the Jewish interpretation of the book. As with the Westermann commentary, some of Sarna's conclusions will not be acceptable to the baraminologist. For example, Sarna equates *min* with species. In general, however, Sarna provides a refreshingly different perspective from the standard Christian interpretations.

4.7. *Putting It All Together*

Having completed the gathering of data on relevant passages, the most important and sometimes most difficult task begins. The baraminologist must draw proper conclusions from the biblical data. At this stage, it is very easy to make critical errors that could damage baraminological studies and theories. Baraminologists need to strive for exegesis of the text, rather than placing a meaning into the text that is not there.

Good exegesis begins with a good understanding of the biblical terminology. If disagreement over the meaning of particular words exists, avoid building an argument solely on one meaning. Frame the interpretation within the meaning of the words present in the text, and if doubt exists, keep the interpretation open or conservative. When faced with variant meanings, try to bolster your understanding of the organisms in question by referring to other passages from the same book where the same or related terms appear. For example, the Hebrew term *bar* means "grain," but it also carries the related meaning of "field." In Genesis 41:49, *bar* refers to grain (KJV: "corn"), but in Job 39:4, it appears to mean "field" instead. The variant meaning has little impact on the study of grasses in the Bible, since Job refers unambiguously to other cereal crops in other passages (31:40; 24:6; 24:24).

Next, examine the context in which the word occurs. Since baraminologists are most interested in historical data, context may have less use than the words themselves. For example, poetical works often use figures of speech, like metaphors or hyperbole, that cannot be properly interpreted from their plain meaning. For example, Job laments, "Oh that my grief were thoroughly weighed, and my calamity laid in the balances together! For now it would be heavier than the sand of the sea" (Job 6:2–3). We ought not take this to mean that poetry cannot contain true descriptions of organisms or history. Job also contains lengthy descriptions of organisms, particularly in the Lord's speeches in Job 38–40. When considering poetry, remember to discriminate between figures of speech and realistic descriptions. Once again, when in doubt, avoid dogmatic assertions.

To determine the historical setting of passages in Genesis, try to place the reference into the genealogy as listed in Genesis 5 and 11. By using this chronology, creationists can estimate the date of nearly any reference in Genesis. Baraminologists may also wish to consult a chart displaying the life spans of the Genesis patriarchs to aid in this process. References in Job can be referred to the biblical chronology by Job's approximate contemporaneity with Abraham.

4.8. Chapter Summary

This chapter focuses exclusively on gathering data from the most reliable source we have–the Bible. The Bible is authoritative and without error. It is the primary source of belief in creation, and, therefore, should be one of the first resources to be analyzed when looking for baramins and studying their history. We should keep in mind a list of relevant questions for our biblical analysis:

1. Is my group mentioned in the Bible?
2. When does my group first appear in the Bible?
3. How did my group survive the Flood?
4. What does the first scriptural appearance and the Flood survival tell me about my baramin?
5. Does my baramin show signs of "groaning and travailing" under the Curse?

In interpreting the Bible, we must be careful to examine the context in which the passage was written. This must be a meticulous examination of both historical as well as literary context. Baraminologists should consult professional Bible scholars, but may do some research on their own using the Bible Gateway, concordances, and commentaries.

Review Questions

1. In what two ways can the Bible aid baraminology research?
2. List five key questions that baraminologists should try to answer from Scripture.
3. What passage tells us that discontinuity existed at Creation?
4. What do camels and grasses tell us about the history of baramins after the Flood?
5. Why is the Book of Job important for baraminology research?
6. How can baraminologists verify translation?
7. What is meant by the "meaning" of a text?
8. How do we accurately find the context in which a section of Scripture is written?
9. What is the danger of interpreting Scripture without investigating context?
10. What is the primary Hebrew text used by modern translators? What text was used for the King James translation?
11. Describe the Septuagint.
12. What information can we gain from Bible commentaries?

13. Define exegesis.
14. Why should baraminologists be careful when handling the Bible? What mistakes could be made?
15. Why should baraminologists compare Bible translations?
16. How can we estimate dates of events described in Genesis?
17. Why is it important to understand the historical context and literary genre of the biblical passages being analyzed?
18. Many passages in Job employ poetry as a writing style. How could poetry complicate the translation and analysis of such passages?
19. Should a baraminologist attempt to translate a complicated or isolated Hebrew term?

For Further Discussion

1. How can biblical arguments backfire on baraminologists? What techniques can we use to avoid such mishaps?
2. What might be some baraminological implications of the mention of the "dove" and "raven" in the Flood story?
3. If you could create an ideal concordance for the "scientific theologian," what would you put in it and why?
4. Can we ever have a perfect translation of the Bible? Why or why not? Is such a translation necessary?
5. How would you go about selecting a good Bible commentary for biological research?
6. Make a list of other possible resources, not mentioned in this chapter, that might aid in the investigation of biblical context.
7. Is there further research that can be done to refine the biblical timeline that modern creationists use? What would be required for such a project?
8. How do scientists identify Bible scholars who are reliable and true to the Genesis text?
9. How do we know whether our baraminology hypotheses fit within the revealed truth of Scripture?
10. Using the Bible Gateway, complete the following chart:

Verse	KJV	NASB	NIV
Lev. 11:19	lapwing		
Deut. 14:5		ibex	

CHAPTER 5

Successive Approximation

5.1. Introduction

The most important advancement of modern baraminology came with the abandonment of hybridization as the sole criterion for identifying baramins. Although Marsh placed much emphasis on hybridization, he often strayed into tautology. For example, Marsh endorsed hybridization as the criterion for baraminic membership while at the same time claiming that no evidence had ever shown hybridization between members of two different baramins.[1] Any hybridization was by Marsh's definition evidence that the parents belonged to the same baramin; therefore, there could be no evidence of hybridization between baramins. We can avoid this tautology by admitting other evidences of baraminic membership.

In the 1980s, ReMine developed the idea that baramins should be approximated by multiple criteria, effectively circumventing Marsh's tautologous baramin. With the admission of additional baraminic membership criteria, baramins could be inferred even when hybridization cannot be applied or when it fails completely. He called this idea "successive approximation" because the membership of the baramin is successively refined through the accumulation of both additive and subtractive evidence. ReMine's admission of additional evidence cleared the way for the development of a morphology-based baramin definition, such as the refined baramin concept.[2]

Central to the practice of successive approximation are the three terms **monobaramin**, **apobaramin**, and **holobaramin**. To review briefly, a *monobaramin* is a group of known organisms or species that share continuity. An *apobaramin* is a group that is discontinuous with all other organisms. A *holobaramin* is a group that is both continuous among all members and discontinuous with all other organisms. Since the holobaramin is simultaneously a monobaramin and an apobaramin, we may think of any apobaramin or monobaramin as an *approximation* of the holobaramin.

Some monobaramins or apobaramins contain nearly the same species as the holobaramin and therefore approximate the holobaramin well. Other monobaramins and apobaramins contain too few or too many members and approximate the holobaramin poorly. As we add more species to a monobaramin, the monobaramin becomes a better approximation of the holobaramin. As we divide an apobaramin, the apobaramin becomes a better approximation of the holobaramin. When the membership of an apobaramin and a monobaramin coincide, a holobaramin has been defined.

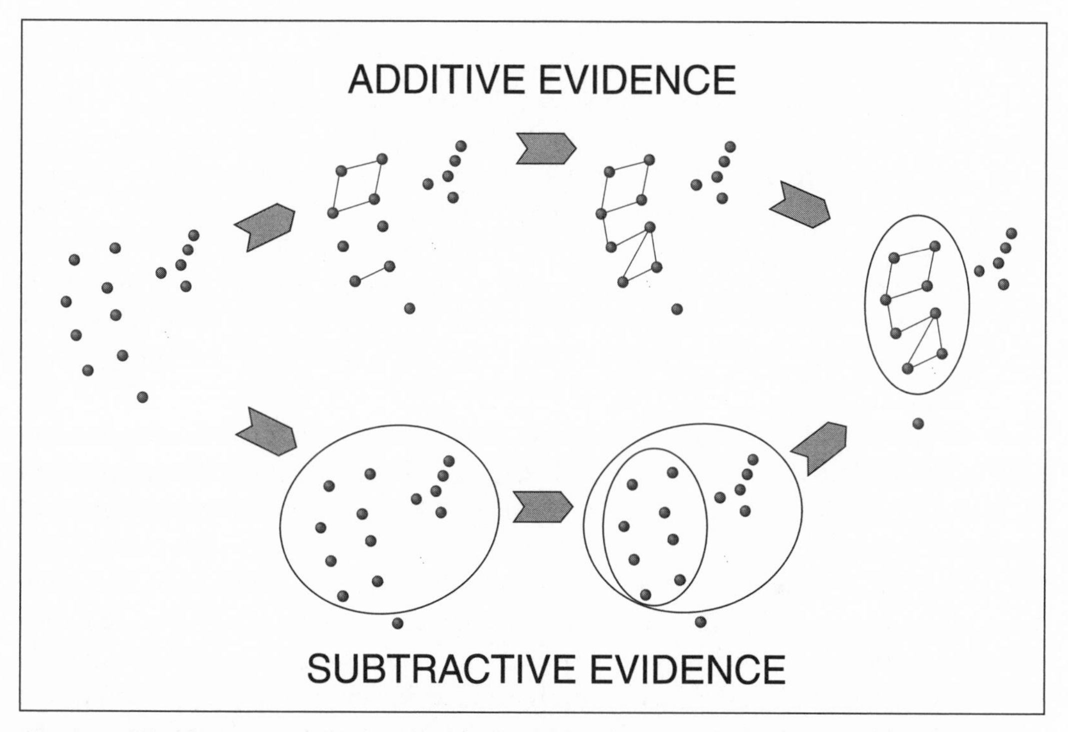

Figure 5.1. *Successive approximation approaches the holobaramin from two different directions. In one direction (top), species (represented by dots) are combined into larger groups by additive evidence (represented by connecting lines). In the other direction (bottom), species are divided into smaller groups bounded by discontinuity (represented by circles). Eventually, the approaches should converge on a group with the same membership, the holobaramin.*

Since the holobaramin and baramin are defined differently, we may think of the holobaramin as an approximation of the baramin. Presumably, as more and more holobaramins are identified, the characteristics of these holobaramins will reveal attributes about the underlying baramins, which in turn will reveal attributes about the underlying potentiality regions. The potentiality regions will then provide a picture of God's plan of biological creation. At this stage, baraminologists have identified very few baraminic groups when compared to the vast array of diversity that God created. A good understanding of baramins and potentiality regions remains a distant goal.

In this chapter, we will present a general overview of the successive approximation strategy, beginning with a discussion of the starting point for baraminological studies and moving to examples published in the technical literature. For the sake of understanding successive approximation, we will explain the methods of identifying continuity and discontinuity in later chapters. For now, we will merely refer to methods that establish continuity and discontinuity.

5.2. Intuition

We can begin baraminology research with our own scientific intuition. The idea of intuition is unacceptable to most scientists, but in creation biology, we can justify its use. Intrinsic to the refined baramin concept is its recognizability. Baramins are not poorly defined, ethereal groups. God created human beings with the inherent capacity to recognize discrete groups in His living creation. Adam's naming of the birds and beasts was in part an expression of his ability to recognize that which God had made. Even today, we readily recognize mammals, reptiles, birds, swimming creatures, and plants, not because we have received formal biological training but because God gave us that power of recognition.

Because God gave us the ability to recognize baramins, baraminologists should not underestimate or ignore the power of their own intuition. Since human beings, not Christians alone, naturally recognize groups of organisms, we may use these groups as the starting point for our baraminology research program. We may even find value in the standard classification scheme, even though it is heavily influenced by phylogenetic research. The recognizability of baramins among standard classification is reflected in the common creationist belief that "families," such as Felidae or Poaceae, approximate the limits of baramins.

Though we recognize our intuition as a starting place, we must not place too much confidence in our intuitive abilities. Linnaeus recognized discrete species but overemphasized their place in creation biology. Later in his life, when faced with the crisis of interspecific hybridization, he introduced logically inconsistent modifications to his species concept. Had he carefully gathered more evidence before asserting that all species remain fixed from the Creator's hand, biology might be very different today.

We must practice theoretical restraint and scientific vigilance to avoid such easily repeated errors. Do not elevate baraminological hypotheses, or even baraminology itself, to the level of authority that God and His Word enjoy. Baraminology is just science, and it will change. We must review the theories of others critically, and hold our own theories loosely. Above all, our faith and allegiance must belong to God, not science.

Baraminologists, like all scientists, seek correspondence of theory with reality. Intuition constitutes part of the reality that we include in our baraminological hypotheses. Like any evidence, however, intuition is insufficient to establish the identity of holobaramins because holobaramins must be identified holistically. Intuition merely serves as the starting point, upon which we add other kinds of additive and subtractive evidence. In this way, the identity of a holobaramin can be firmly established.

5.3. An Example of Successive Approximation

Having identified an intuitive group, we must then seek evidence to confirm or deny the baraminological status of the group. For example, we recognize the group "birds," but birds certainly consist of more than one holobaramin. Thus, the intuitive group "birds" furnishes an excellent foundation for future study that subdivides the group. In contrast, tigers also form a well-defined, discrete group. Rather than being subdivided, however, tigers may be grouped with other cats based on evidence of continuity. Here, we will discuss the cat family Felidae in more detail, as an example of the successive approximation approach.[3]

The cat family Felidae contains thirty-eight species in six genera, with such well-known members as the lion, tiger, jaguar, cheetah, puma, and the common house cat. Like many other mammalian families, the cats first appear in the post-Flood fossil record and subsequently diversify rapidly into many different species. Some species, such as the sabertoothed cats, became extinct during the Decimation. Today, wild cats inhabit every continent except Australia and Antarctica. Modern domestic cats possibly grew tame by repeated exposure to humans through their hunting of rodents so often found in agricultural settlements.

To evaluate the baraminic status of the Felidae, Robinson and Cavanaugh examined a holistic set of data derived from ecological, morphological, chromosomal, and molecular information. They selected seventeen of the thirty-eight extant species to study for evidence of within-group continuity. They also included the spotted hyena (*Crocuta crocuta*) and the meerkat (*Suricata suricatta*) as outgroup comparisons to detect discontinuity. They selected the outgroup species based on their classification into a single superfamily with the cats, reasoning that these two species are the most likely to show continuity with the cats, if any continuity might exist.

In evaluating evidence for continuity, Robinson and Cavanaugh found that the seventeen species in their study formed three strongly continuous groups (monobaramins). The big cats (lion, leopard, jaguar, tiger, snow leopard, and clouded leopard) formed one monobaramin, the cheetah and puma formed another, and the small cats (everything from the domestic cat to the bobcat and ocelot) formed the third. At this stage, it would be tempting for the baraminologist to stop and conclude that the cats comprised three different baramins, but the evidence of continuity does not warrant that conclusion yet. Evidence of discontinuity must be sought in order to identify any group other than the monobaramin.

As Robinson and Cavanaugh extended their study of continuity among the felids, they found several important facts. They first found that the puma and the jaguar could hybridize, indicating continuity between the species (see chap. 7 for discussion of the hybridization criterion). Since the puma and jaguar belong to different monobaramins, the continuity between the species established probable continuity between the monobaramins also. Further study of the baraminic distances between the cat monobaramins (see chap. 8) confirmed the existence of continuity between all cat monobaramins. Consequently, they concluded that all seventeen cat species that they examined belonged to one large monobaramin.

Even though we can classify the family Felidae as one monobaramin, its status as a holobaramin remains inconclusive until evidence of discontinuity between the cats and other species can be demonstrated. To do this, Robinson and Cavanaugh studied the similarities between cats, hyenas, and meerkats. Cats showed significant dissimilarity with these two outgroup species. Since other carnivores and mammals are even more different from the cats than hyenas and meerkats, Robinson and Cavanaugh concluded that the family Felidae is an apobaramin, set apart from all other species by clear and detectable discontinuities. The fossil record of the cats supports their conclusion because clear transitional species linking cats to other carnivores have yet to be discovered.

Having demonstrated continuity among the cats and discontinuity between the cats and other species, Robinson and Cavanaugh concluded that the cats form a holobaramin. Still, their conclusion can only be regarded as a tentative hypothesis. At each step of their research they made assumptions about the data to simplify the analysis. Each of these assumptions was both reasonable and warranted; nevertheless, one or more of their assumptions could prove incorrect. They assumed that the seventeen species of cats they examined represented the full variety of extant cats. Because they selected a good diversity of species, this assumption is reasonable.

They also assumed that they only needed to examine outgroup species that are most similar to the cats, reasoning that if discontinuity between cats and the species most similar to them could be established, then discontinuity would certainly exist between cats and more dissimilar species. If these assumptions prove incorrect, it is possible (though unlikely) that a divergent member of the cat family might belong to a different baramin. It is also possible (but also unlikely) that an unusual non-Felid carnivore might prove to be a member of the cat holobaramin.

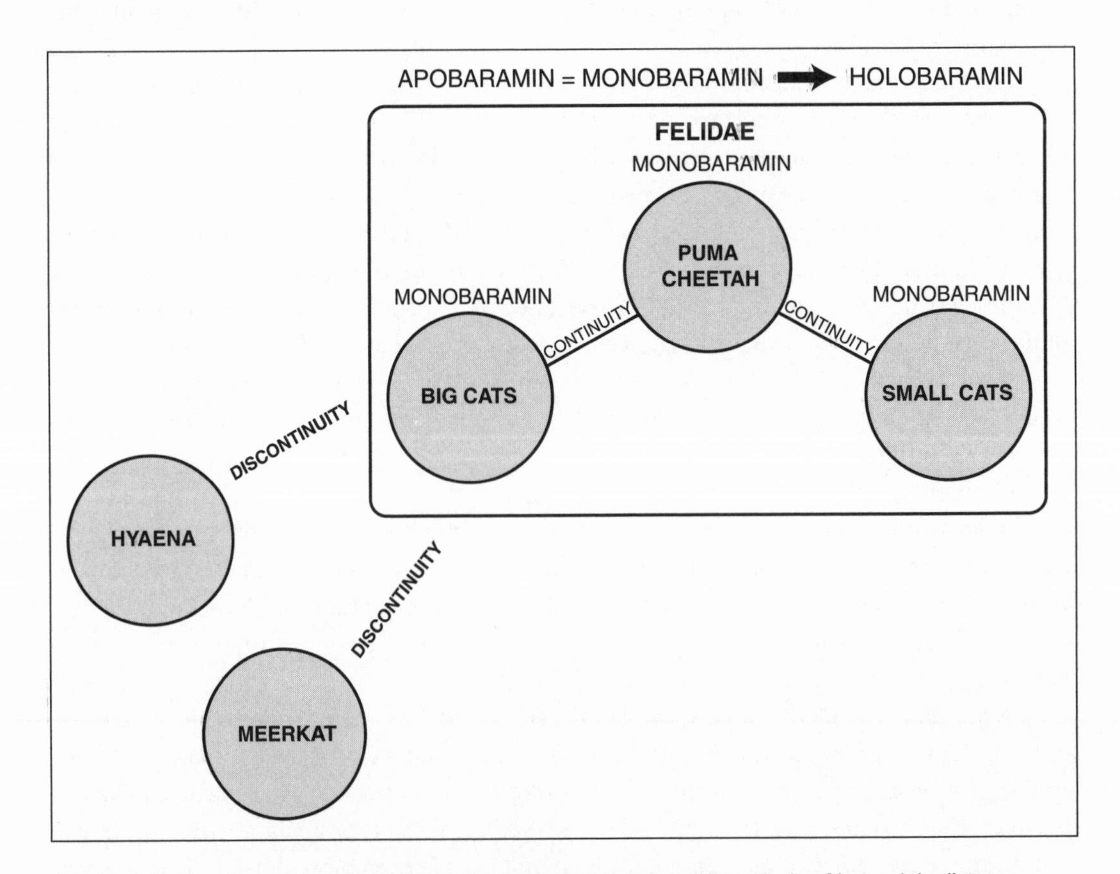

Figure 5.2. *Monobaramins of the Felidae can be connected through analysis of baraminic distance.*

5.4. Context

The crucial feature that separates baraminology from wishful thinking is context. Theoretically, continuity or discontinuity can be detected at any taxonomic level. When compared to bacteria or plants, meerkats and cats appear to share significant continuity. Similarly, if we examined the variation of a hundred individual tigers, house cats would appear discontinuous. As our focus or context changes, the significance of similarity or dissimilarity changes also. How do baraminologists maintain even a semblance of scientific credibility when continuity and discontinuity shift so easily? In other words, what is the proper context for detecting continuity and discontinuity?

The answer to this question depends greatly on your own point of view. Evolutionists find all of baraminology an exercise in absurdity and view this apparent lack of objectivity as evidence that baraminology is not science. We creationists rest instead on the philosophical and biblical foundation explained in chapter 2. Since we believe that something like a "discrete unit of biological creation" (to use a Marshian term) must exist, detecting baramins becomes a matter of adjusting our context until the baraminic limits emerge.

To a creationist, detecting baramins bears a strong resemblance to looking at microorganisms under a microscope. At the wrong magnification or with the wrong focus, nothing can be seen in the microscope field, but when the focus and magnification converge on the appropriate settings, the microbes come into sharp focus. The baraminologist seeks the proper focus for observing baramins because baraminologists believe there is something real to observe. In contrast, the evolutionist rejects the very existence of baramins and finds our focusing efforts nonsensical and arbitrary.

Baraminologists find proper baraminic context by considering four lines of evidence. First, intuition provides a framework in which baraminological hypotheses may be constructed and tested. Second, the Bible offers some guidelines in the creation account. Based on the separate creation of vertebrates at approximately the level of class as discussed in chapter 4, we may assume that classes contain more than one baramin. Thus, baramins are most likely near the order or family level of classification. Whether or not this trend is generally applicable to non-vertebrates, its usefulness with the vertebrates makes it at least worthy of consideration for other groups. Third, the short history of the earth would seem to preclude megaevolutionary events, such as the origin of new phyla or classes during only six thousand years since creation. Fourth, the post-Flood fossil record frequently shows sudden appearances of families, followed by their subsequent diversification. Taking all four of these considerations together, the baramin may be generally equated with an order, family, or tribe, rarely with something broader or narrower.

Returning to the microscope analogy, the microscopist generally knows the approximate size of the organisms or structures sought and adjusts the microscope settings accordingly. In the same way, by setting our focus at the family or order level of classification, baraminologists should find evidence of continuity and discontinuity that demarcates holobaramins. Even with the proper microscope settings, organisms of interest are rarely in perfect focus at first glance. Similarly, when continuity or discontinuity cannot be detected, baraminologists should adjust their focus slightly broader (e.g., from tribe to family) or narrower (e.g., from order to family) as circumstances dictate.

To someone who believes the microscope slide is empty, focusing is a waste of time, but to a skilled microscopist who believes the slide contains visible life, focusing is a legitimate activity, even though it seems random and arbitrary. The microscopist knows that focusing works because he knows microorganisms when he sees them. Likewise in baraminology, adjustment of context is not arbitrary because we have good reasons to believe that baramins exist and that baramins are approximately as large as orders or families. Like the microscopist recognizing microbes, skilled and experienced baraminologists will know baramins when they see them.

5.5. The Importance of Apobaramins and Monobaramins

Many creationists have a very absolutist perspective on baraminology, partly stemming from Marsh's emphasis on interspecific hybridization as the only defining characteristic of baramins. Creationists still seek a characteristic that will work for all organisms and will unambiguously identify the "created kinds." Baraminology eliminates this absolutist perspective, with successive approximation as the alternative strategy. If we could only identify apobaramins or monobaramins as absolutists, we would fail because our objective is the holobaramin. By abandoning the absolutist position, baraminologists succeed when detecting continuity or discontinuity because each monobaramin and apobaramin is an approximation of the holobaramin. Successive approximation infuses apobaramins and monobaramins with value and utility not as failed attempts to identify holobaramins but as meaningful approximations of holobaramins.

In fact, the majority of published baraminology studies identify only monobaramins or apobaramins rather than holobaramins. Since hybridization is additive evidence, all of the basic types identified by Siegfried Scherer's colleagues constitute monobaramins (see chap. 2).[4] Wise's first application of baraminology identified the turtles as an apobaramin.[5] Wood and Cavanaugh analyzed plants from the sunflower family and found a monobaramin.[6] All of these studies provide valuable baraminological evidence and should not be discounted merely because they failed to identify a holobaramin. As more evidence and newer methods become available, baraminologists will further refine these holobaraminic approximations. In the meantime, baraminologists can continue to identify more monobaramins and apobaramins.

5.5.1. Sunflowers

The sunflower family Asteraceae is one of the largest plant families, with approximately twenty thousand species.[7] Because this level of diversity is unknown in the vertebrates, Wood and Cavanaugh began to study sunflower baraminology at the level of a single subtribe, Flaveriinae, commonly called yellowtops. All scholars agree that Flaveriinae contains at least three genera, *Flaveria, Sartwellia,* and *Haploësthes,* twenty-eight species in all. Based on hybridization, all twenty-eight species showed continuity, indicating their membership in a common monobaramin. By analyzing a morphological dataset with a multidimensional pattern recognition technique (see chap. 9), Wood and Cavanaugh also discovered continuity between the Flaveriinae and other sunflower family members, including marigolds. Based on their results, they were unable to detect any discontinuity and hypothesized that the Flaveriinae constituted a part of a larger holobaramin.

In a follow-up study, the same research team evaluated a much larger morphological dataset with more taxa and more characters, using the same pattern recognition technique. The new dataset represented some 5,730 species from 3 different tribes (Heliantheae, Helenieae, and Eupatorieae), including such species as sunflowers, marigolds, zinnias, ragweeds, and black-eyed Susans. Pattern recognition analysis revealed two continuous populations that did not correspond to any of the conventional tribal or subtribal classifications. Based on their analysis, they concluded that all of the species were members of a single monobaramin. Once again, their research revealed no discontinuity between any of the organisms studied.

In the same study, Cavanaugh and Wood presented limited evidence suggesting that the sunflower family constitutes an apobaramin, discontinuous with all other species. They based their claim of discontinuity on the presence of a unique suite of characteristics defining the family Asteraceae and on the fossil record, which exhibits a full range of diversity in the family at its first appearance. These considerations may indicate discontinuity, but a fuller examination of the evidence is necessary before true discontinuity can be claimed (see chap. 6).

The studies of the Asteraceae illustrate the method of successive approximation well. The researchers began with a small group and established continuity with other members of the family. They expanded their focus and found even more evidence of continuity with

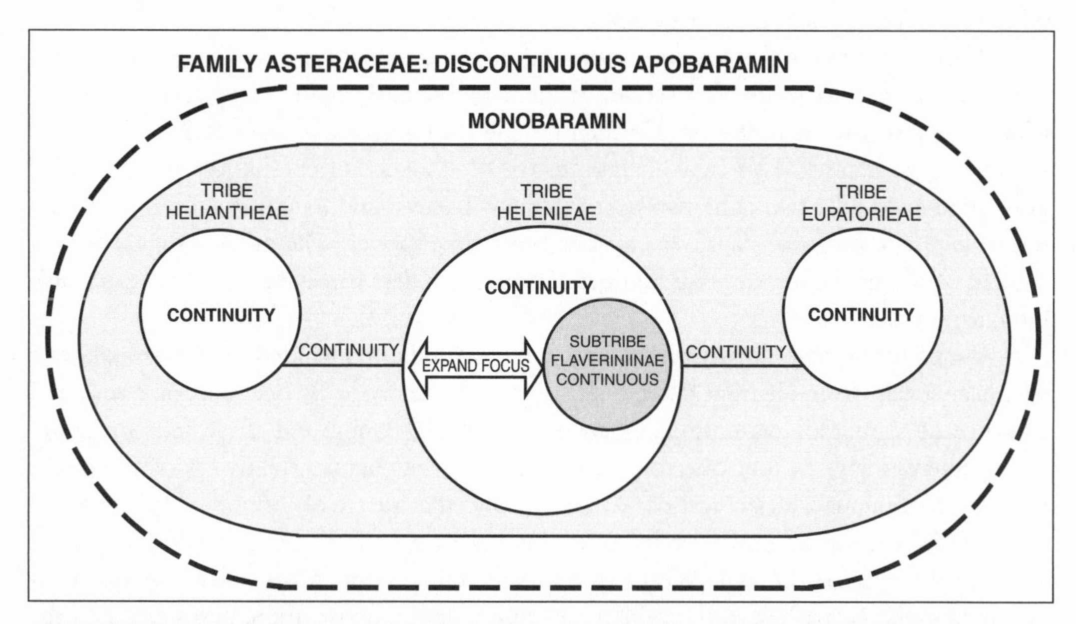

__Figure 5.3.__ Baraminological research in the flower family Asteraceae began with a small monobaramin corresponding to the subtribe Flaveriinae. Further evidence led to the classification of the entire tribe Heliantheae (of which Flaveriinae is a member) as a monobaramin. Limited evidence suggests that Asteraceae is an apobaramin.

members of three different tribes. At the same time they tried to examine the Asteraceae family for evidence of discontinuity. To continue studying the baraminology of the sunflower family, future researchers will need to broaden their perspective to include all of the twenty thousand species of the family to detect potential discontinuity with other plant families. Alternatively, an additive approach may be used, and any of the other twenty Asteraceae tribes may be studied for evidence of continuity and discontinuity. In this way, the baraminological status of the Asteraceae can be firmly established with multiple scientific studies, each of which successively refines previous approximations of the holobaramin.

5.5.2. Wheats

Many of the most important cereal grains in the world are members of the grass tribe Triticeae. Although undergoing constant revision, the species count for Triticeae numbers around 360, including the economically important ryes, barleys, and wheats.[8] Junker evaluated interspecific and intergeneric hybridization among the Triticeae species and found that hybridization is very common in this group. Based on these results, he concluded that the entire tribe formed a basic type or monobaramin.

In a broader follow-up study, Wood examined evidences of continuity and discontinuity among all 10,000 species of the grass family Poaceae. He first evaluated evidences of intertribal hybridization among the grasses and found many examples of hybridization between members of different tribes. Beginning with a survey of hybridization evidence, he proposed that approximately 7,200 grass species, including the entire tribe Triticeae, belonged to a single monobaramin. Members of the Triticeae monobaramin hybridized with members of ten other tribes, conclusively demonstrating that Triticeae shares continuity with other grasses and is not an apobaramin.

Wood's survey of grass hybridization left the baraminic position of several groups undefined. In particular, the 825 species of the bamboo tribe could not be related to other tribes by intertribal hybridization. He then broadened his focus by statistically analyzing a morphological dataset. This new analysis provided evidence of continuity between every grass tribe except for two. The two tribes (Streptochaeteae and Anomochloeae) that could not be joined to the larger family comprised only three species. The other ~10,000 species showed continuity, demonstrating that the entire family (less three species) formed a single monobaramin.

Along with representatives of the grass family, Wood also included four representatives of similar species from different families in his statistical analysis. By doing so, he discovered evidence of significant discontinuity between the grass family and these four outgroup plants. Interestingly, the four outgroup species and the members of the two grass tribes that were not continuous with the rest of the family showed evidence of continuity among themselves. Although it would be tempting to evaluate the continuity among the outgroups, the focus on the Poaceae probably means that the baraminic status of the outgroup species is poorly focused. Nevertheless, the evidence of discontinuity between the outgroups and the rest of the grass family establishes the apobaraminic status of most of the grass family.

Since the grass family demonstrates continuity among most of the tribes and discontinuity between the same tribes and other species, we can propose that most of the family

Poaceae represents a single holobaramin. As an extension of Junker's earlier study of the tribe Triticeae establishing its monobaraminic status, Wood's research found additional baraminic evidence by broadening the baraminological focus. Future studies may confirm or deny the holobaraminic status of the grass family, including further evaluations of the two tribes that do not show continuity with other members of the grass holobaramin. Future refinement of the grass holobaramin will firmly establish its reality.

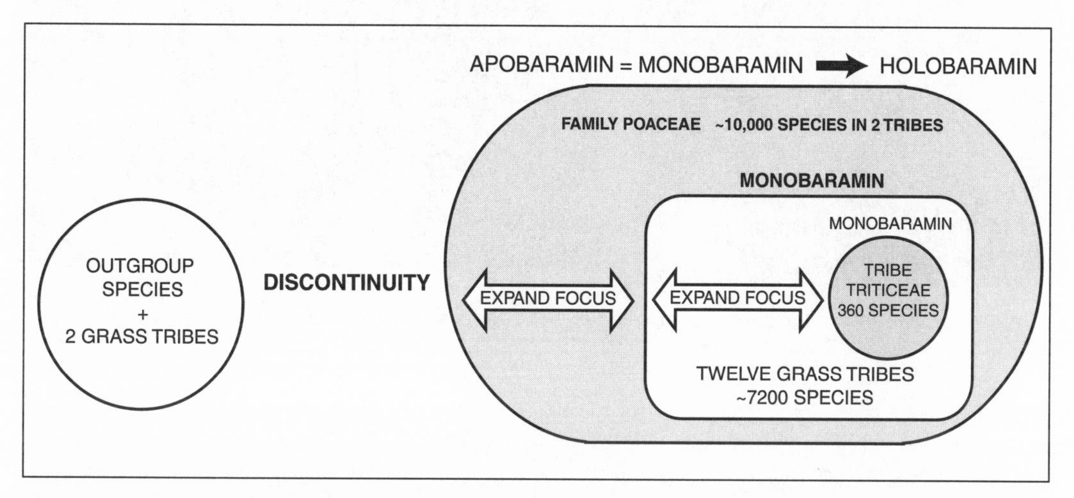

Figure 5.4. *Baraminological research in the grass family Poaceae began with the wheat tribe Triticeae classified as a monobaramin. Subsequent research showed that members of other grass tribes could hybridize with members of Triticeae, expanding that monobaramin to include approximately 7,200 species. Further morphological analysis revealed a discontinuity around nearly the entire grass family and continuity among most of its members. Consequently, the family Poaceae is probably a holobaramin.*

5.5.3. Turtles

By far the most studied baraminic group in all of creationism, the turtles present a different perspective on successive approximation because discontinuity is easily established but evidence of continuity has thus far been more elusive. Scientists classify turtles into the order Testidunes, consisting of 13 families and 260 species. Turtle species live in both tropical and temperate climates, and some marine forms have been found in Arctic waters. Unlike mammalian baramins, the turtle fossil record begins in the Flood sediments, with gargantuan specimens preserved in the uppermost layers of Flood rocks. Wayne Frair, a founding member of the Creation Research Society, spent thirty years actively researching the turtles and has written several papers on their baraminology.

By examining turtle blood proteins, Frair found some evidence of continuity among a number of different varieties of turtles. For example, he discovered that all the sea turtles showed a high degree of similarity. He also proposed that the European pond turtle (*Emys orbicularis*) and the tortoises shared a common ancestor. Frair relied heavily on the absence of a widely accepted transitional form between turtles and other reptiles as evidence that turtles are discontinuous with other living things. Frair regarded the fossil turtle *Proganochelys* as the ancestral form for all modern turtles, but did not strongly justify this claim. In an earlier study, Frair suggested that the turtles were composed of four different baramins, and he did

Figure 5.5. *The gopher tortoise, snapping turtle, and sea turtle are members of order Testudines.*

not rule this option out in his later work. We can summarize Frair's model of turtle baraminology as accepting the apobaraminic status of turtles and proposing at least four monobaramins within the apobaramin.[9]

In a later study, Wise surveyed various evidences of discontinuity to detect the apobaraminic status of the turtles and of various groupings within the turtles. He used a battery of characteristics to demonstrate convincingly that discontinuity exists between the turtles and other organisms. In addition, he also suggested that some evidence might support the division of turtles into two smaller apobaramins, the pleurodires (the so-called "side-necks") and the cryptodires ("straight-necks"), and the splitting of turtles into four apobaramins equivalent to the monobaramins suggested by Frair in his earlier study. Wise's study contributed to the refinement of turtle baraminology, but further research would be necessary to more clearly delineate the baraminic boundaries of turtles. Most importantly, because Wise looked at no evidence of continuity, we cannot draw from his work evidence for holobaramins.[10]

Most recently, Robinson examined turtle baraminology using mitochondrial DNA (mtDNA). He provided four observations that confirmed the discontinuity between turtles and other organisms. With Robinson's reliance on DNA for evidence of discontinuity, he expanded the holistic argument for the turtle apobaramin beyond the morphological evidence used by Wise and Frair. Using DNA, Robinson confirmed the discontinuities detected by morphological studies. As a result, the turtle apobaramin may be one of the best supported apobaramins in the history of baraminology research.[11]

Robinson also examined additive evidence for turtle monobaramins. He began by comparing the mtDNA similarity between species of turtles that could successfully hybridize. Using this method, he was able to detect two turtle monobaramins, the turtle family Cheloniidae and the genus *Gopherus*. Because no other hybridizing species had mtDNA sequences available in the public database, Robinson refrained from further speculation on the monobaraminic nature of the turtles. In summary, Robinson provided further evidence for the apobaraminic status of turtles and began the task of building monobaramins based on holistic evidence (hybridization and mtDNA similarity).

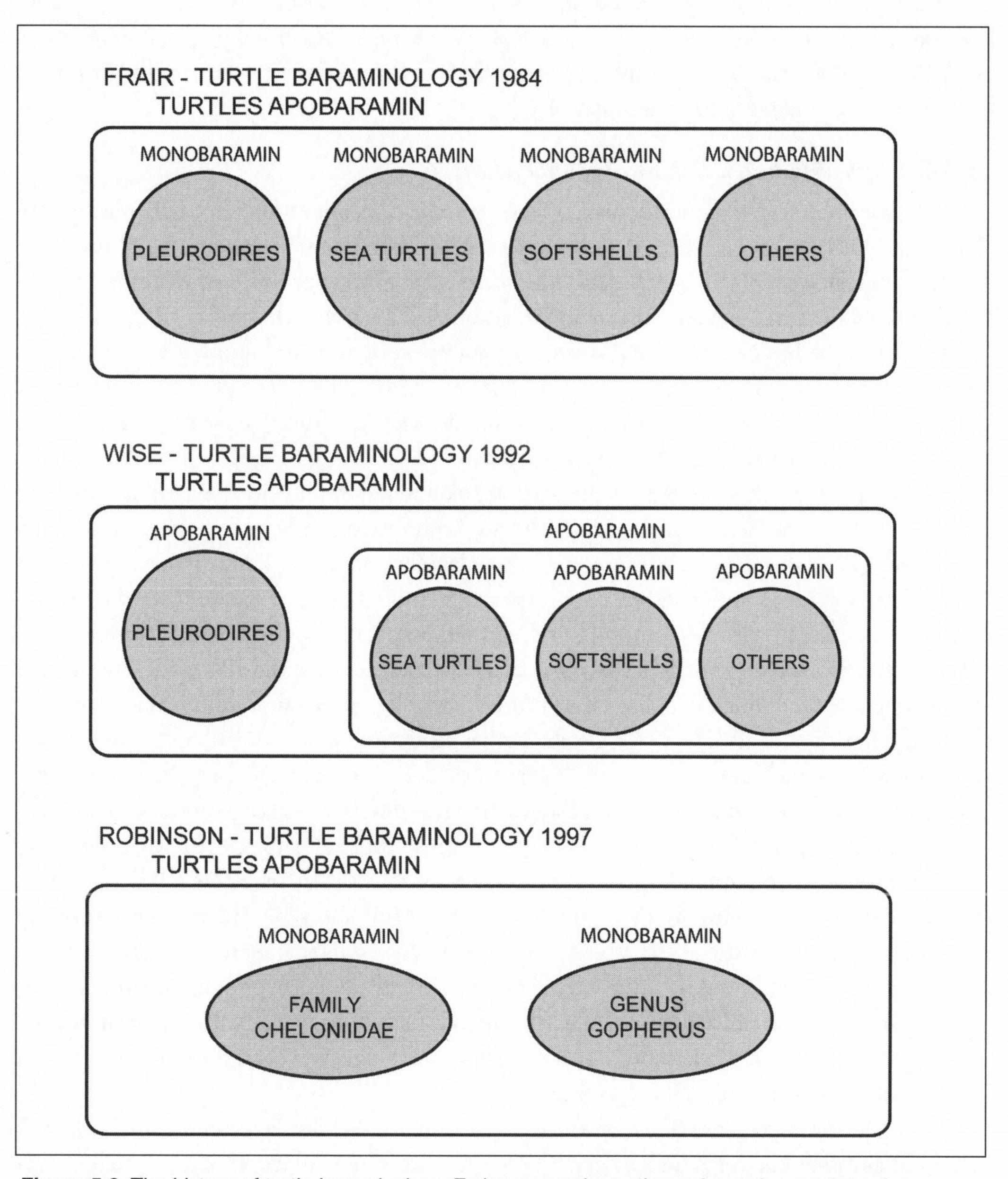

Figure 5.6. *The history of turtle baraminology. Frair proposed a turtle apobaramin containing four monobaramins. Wise proposed a turtle apobaramin that could possibly be subdivided into other apobaramins. Robinson took a more conservative approach. He agreed that the turtles were an apobaramin but proposed only two small monobaramins. Much work in turtle baraminology remains to be done.*

To further refine turtle baraminology, baraminologists can begin with the common hypothesis of Frair and Wise that turtles comprise four holobaramins: the pleurodires, the sea turtles, the softshells, and the other cryptodires. Frair has provided good evidence based on blood proteins that these groups represent monobaramins, but more evidence is needed before we can claim that continuity exists within these groups. Wise's approach to detecting discontinuities should be expanded to include statistical studies of turtle morphology before his four apobaramins are widely accepted. Robinson's approach of building monobaramins species by species can also be employed, since the small size of the turtles makes baraminological research manageable and practical.

5.5.4. Other Apobaramins and Monobaramins

As baraminologists continue to study God's creation, more apobaramins and monobaramins will be proposed. In this section, we will present a brief review of two of the latest groups to appear in the technical literature. The first group, salamander family Plethodontidae, is the first attempt to apply baraminology to an amphibian group. The second study, of the bacteria of family Mycoplasmataceae, is a more ambitious effort to identify baraminic groupings in asexual organisms. The refined baramin concept applies to bacteria and archaea too; therefore, baraminologists should seek methods to identify baramins in these groups, challenging though they may be.

The family Plethodontidae is the largest salamander family in the world, with 240 species in 27 genera. The majority of species are found exclusively in the New World, with the greatest diversity found in the southern Appalachian mountain range. Every member of the family lacks lungs and breathes only through the moist skin and the lining of the mouth. Because of their complex mating rituals, interspecific hybridization among the Plethodontidae rarely happens. Based primarily on morphology, development, and fossils, Wood hypothesized that the entire family constitutes one apobaramin and is discontinuous with all other organisms. His work awaits further study to confirm or refine his hypothesis. Future studies in the salamanders could focus on gathering a comprehensive dataset to establish continuity or discontinuity within the family and to further refine Wood's hypothesis.

Asexual organisms represent some of the most difficult subjects for baraminology because historically baraminologists have relied almost exclusively on interspecific hybridization as baraminic evidence. With no possibility of hybridization, organisms that reproduce asexually must be studied in some other way to establish their baraminic status. Furthermore, the wide variety of information available from a morphological study is also difficult to derive for unicellular organisms. Instead, data from ecology, biochemical metabolism, and genomic similarity could prove more valuable for asexual, unicellular organisms than for their sexual, multicellular counterparts.

The family Mycoplasmataceae consists of tiny bacteria that lack the normal cell wall found in all other bacteria and archaea. Commonly known as mycoplasmas, these bacteria also require sterol for growth and use a different genetic code than the code used by most other organisms. The family includes a number of pathogens, such as *Mycoplasma pneumoniae*, the causative agent of walking pneumonia. Because of their medical importance, scientists have studied the mycoplasmas for decades, and much information about their biology is readily available. In fact, four of the species in the family–*M. genitalium, M. pneumoniae, M. pulmonis,* and *Ureaplasma urealyticum*–have had their genomes completely sequenced.

In his review of the genomics of the mycoplasmas, Wood interpreted the unique genetic code and the lack of a cell wall as evidence for discontinuity with other bacteria. Taken by itself, we could interpret the alternative genetic code as holistic data, since the synthesis of every protein in the genome depends on the ability to decode the genes correctly. Since only one codon differs between the standard code and the mycoplasma code, the force of this holistic evidence may be weakened accordingly. Once again, however, we see that this hypothesis of discontinuity provides baraminologists with a starting point for evaluating the baraminology of these bacteria. As noted above, further confirmation of the discontinuity of mycoplasmas may be derived from studies of ecology or metabolism. In doing so, baraminologists will strengthen or refine the argument for discontinuity in these bacteria.

As evidence for continuity, Wood presented the genomic similarity between *M. genitalium* and *M. pneumoniae.* Copies of every gene of the smaller bacterium, *M. genitalium* with 466 protein-coding genes, are also found in the genome of the larger bacterium, *M. pneumoniae* with 677 protein-coding genes. If created separately, near genomic identity would be very surprising after 6,000 years of divergence. Wood advocates a simpler hypothesis that *M. genitalium* and *M. pneumoniae* recently diverged from a common ancestor. In terms of continuity, the sharing of every gene in the genome certainly counts as holistic evidence, and we may readily accept the monobaraminic status of these two mycoplasmas.

The genome sequences of *M. pulmonis* and *U. urealyticum* do not show the same level of similarity with each other or with the other two mycoplasma genomes. Consequently, the argument used to establish the monobaraminic status of *M. genitalium* and *M. pneumoniae* cannot be used here. Instead, baraminologists need to establish other methods to test for continuity among the mycoplasma species. A better understanding of the dynamics of genomes will also help us to interpret better the evidence of genomic similarity within a baraminological framework.

5.6. Chapter Summary

The most important advancement of modern baraminology came with the abandonment of hybridization as the sole criterion for identifying baramins. In the 1980s, ReMine developed the idea that baramins should be approximated by multiple criteria. He called this idea "successive approximation" because the membership of the baramin is successively refined through the accumulation of both additive and subtractive evidence. Central to the practice of successive approximation are the three terms *monobaramin, apobaramin,* and *holobaramin,* discussed in chapter 2.

The idea of intuition is unacceptable to most scientists, but in creation biology, we can justify its use. Intrinsic to the refined baramin concept is its recognizability. Because God gave us the ability to recognize baramins, baraminologists should not underestimate or ignore the power of their own intuition. Although we recognize our intuition as a starting place, we must not place too much confidence in our intuitive abilities. We must practice theoretical restraint and scientific vigilance to avoid easily repeated errors.

Review Questions

1. What did Marsh believe hybridization to be evidence of?
2. How did ReMine propose that holobaramins should be identified?
3. What is successive approximation? How does it work?
4. What part does intuition play in baraminology?
5. Why is context important to baraminology?
6. What are the four lines of evidence for which baraminologists can find proper baraminic context?
7. Why is interspecific hybridization so rare among Plethodontid salamanders?
8. Why have asexual organisms been so ignored in creationist systematics?
9. List data that might be useful in studying the baraminology of asexual organisms.
10. Give evidence of the discontinuity surrounding the family Mycoplasmataceae.
11. Give two reasons why identification of monobaramins and apobaramins is not a failure.
12. What apobaramin do Wise, Frair, and Robinson agree on?
13. What evidence of continuity did Wise consider in his study of turtle baramins?
14. What distinguishes continuity from similarity?
15. Why are researchers more likely to find apobaramins and monobaramins than they are to find holobaramins?
16. Use the microscope analogy to respond to the criticism that baraminology is arbitrary.
17. What evidence suggests that the small cats and the puma are continuous?

For Further Discussion

1. Could we discover significant similarity between members of different baramins? Why or why not?
2. Why is continuity and discontinuity so reliant upon our points of view? Is this merely a case of bias at work? What is the proper context for detecting continuity and discontinuity?
3. How would you go about refining a holobaramin approximation that includes evidence of both continuity within the group and discontinuity with everything else?
4. What could be done to simplify the baraminology of unicellular organisms?
5. Suppose that all of God's creations were in stasis and that we had identified all possible monobaramins and apobaramins. Would it be easier to identify holobaramins? Why or why not?
6. In baraminology, how is a genome better than a gene?
7. Is it possible to describe all apobaramins and monobaramins?
8. If abandoning hybridization as the sole criterion for identifying baramins was the most important advancement in modern baraminology, what might the next advancement be?
9. What could the baraminology of the Mycoplasmataceae tell us about bacterial baraminology?

CHAPTER 6

Identifying True Discontinuity

6.1. Introduction

The baraminologist's acknowledgment of true discontinuity contrasts starkly with the universal common ancestry of evolution. Despite accepting speciation within baramins, Marsh maintained that discontinuity was the most obvious pattern in the living world. Today, discontinuity is a foundational concept in ReMine's discontinuity systematics and Wise's baraminology. Wise illustrated discontinuity with the analogy of an orchard. Linnaeus's view of species fixity may be illustrated by a lawn, in which each blade of grass represents an unchanging species with its root at creation and its tip in the present. The evolutionary perspective is like a tree, with one root representing the universal ancestor of all living things and the twigs and leaves representing modern species. According to Wise, the view of the baraminologist is like an orchard, with each tree (baramin) having multiple branches (species) but separate roots (origins).[1]

Unfortunately, one of the most common mistakes made by creationists is assuming discontinuity where there is none. To continue the orchard analogy, creationists often discover a branch (species) that appears isolated and assume that it is a tree (baramin) unto itself without examining the branch carefully to find the connection to the larger tree. Linnaeus set the precedent for this habit with his essentialist species concept, and even today, many creationists still accept Linnaean species fixity. Just as accepting too much continuity leads to an incorrect view of the pattern of life (evolution), accepting too much discontinuity also leads to problems and inconsistencies. We must strive to discover the balance between continuity and discontinuity that best represents God's creation.

To review, *discontinuity* is defined as "a significant difference between two sets of organisms detected in a holistic analysis." In some cases, discontinuity can be detected in what appear to be only a few traits, if those traits actually encompass a whole host of characteristics. For example, utilization of a variant genetic code rather than the standard code can indicate discontinuity, as we explained in chapter 2. Because the variant genetic code affects the coding of proteins, which in turn affects the metabolism, which in turn affects the life of the organism, the genetic code has far-reaching implications for the biology of an organism. In general, however, reliance on single traits to define discontinuity should be avoided, as we will see in the following examples.

6.2. The Appearance of "Discontinuous" Organisms in Continuous Groups

Although God's biological design displays a staggering complexity, sometimes creationists too quickly infer unmediated design implementation in spite of evidence pointing to a different conclusion.[2] We often assume that an organism that possesses a complicated trait acquired that trait directly from the Creator. Thus, the presence of the designed trait is used to infer discontinuity (between those that have the trait and those that do not). The complexity of photosynthesis illustrates the danger of this reasoning.

Photosynthesis in flowering plants falls into one of three broad categories: C_3, CAM, or C_4. In C_3 photosynthesis, carbon dioxide is converted to a three-carbon sugar by the enzyme ribulose-bisphosphate carboxylase/oxygenase (rubisco). This three-carbon sugar then goes through a series of reactions called the Calvin cycle, during which the energy from sunlight is stored in the chemical bonds of six-carbon sugar molecules. CAM (Crassulacean Acid Metabolism) photosynthesis occurs most frequently in desert and succulent plants. In CAM photosynthesis, carbon dioxide is stored in four-carbon sugars during the night, then delivered to the Calvin cycle during the day. The stomata of the CAM plant open at night to assimilate carbon dioxide but remain closed during the day to prevent water loss in the desert environment.

C_4 plants exhibit a mechanism of storing carbon dioxide that bears some similarity to CAM photosynthesis. Both CAM and C_4 plants acquire carbon dioxide from the atmosphere and store it in four-carbon sugars. Whereas CAM plants segregate carbon dioxide assimilation from the Calvin cycle by a circadian rhythm, C_4 plants segregate them physically in different areas of their leaves. Within any leaf of a C_4 plant, mesophyll cells comprise the bulk of the leaf material and serve as the site of carbon dioxide assimilation. In C_3 plants, the Calvin cycle takes place in the same mesophyll cells in which the carbon dioxide is acquired, but in C_4 plants, the Calvin cycle only occurs in bundle sheath cells that surround the veins of the leaves. By separating carbon dioxide assimilation from the Calvin cycle, C_4 plants prevent the accidental input of oxygen into the Calvin cycle, which ultimately wastes energy rather than storing it.

At first glance, CAM and C_4 photosynthesis appear much more complex than C_3, and hence might inspire us to infer God's direct design to explain the origin of these variant photosynthetic pathways. It might be tempting for baraminologists to interpret the presence or absence of C_4 photosynthesis as evidence of discontinuity. Upon closer examination, we find precisely the opposite: strong evidence for continuity.

In the sunflower family, the genus *Flaveria* contains species that photosynthesize by the C_3 and C_4 pathways, and even some species that photosynthesize by an intermediate between C_3 and C_4. The existence of intermediates should alert the baraminologist to the

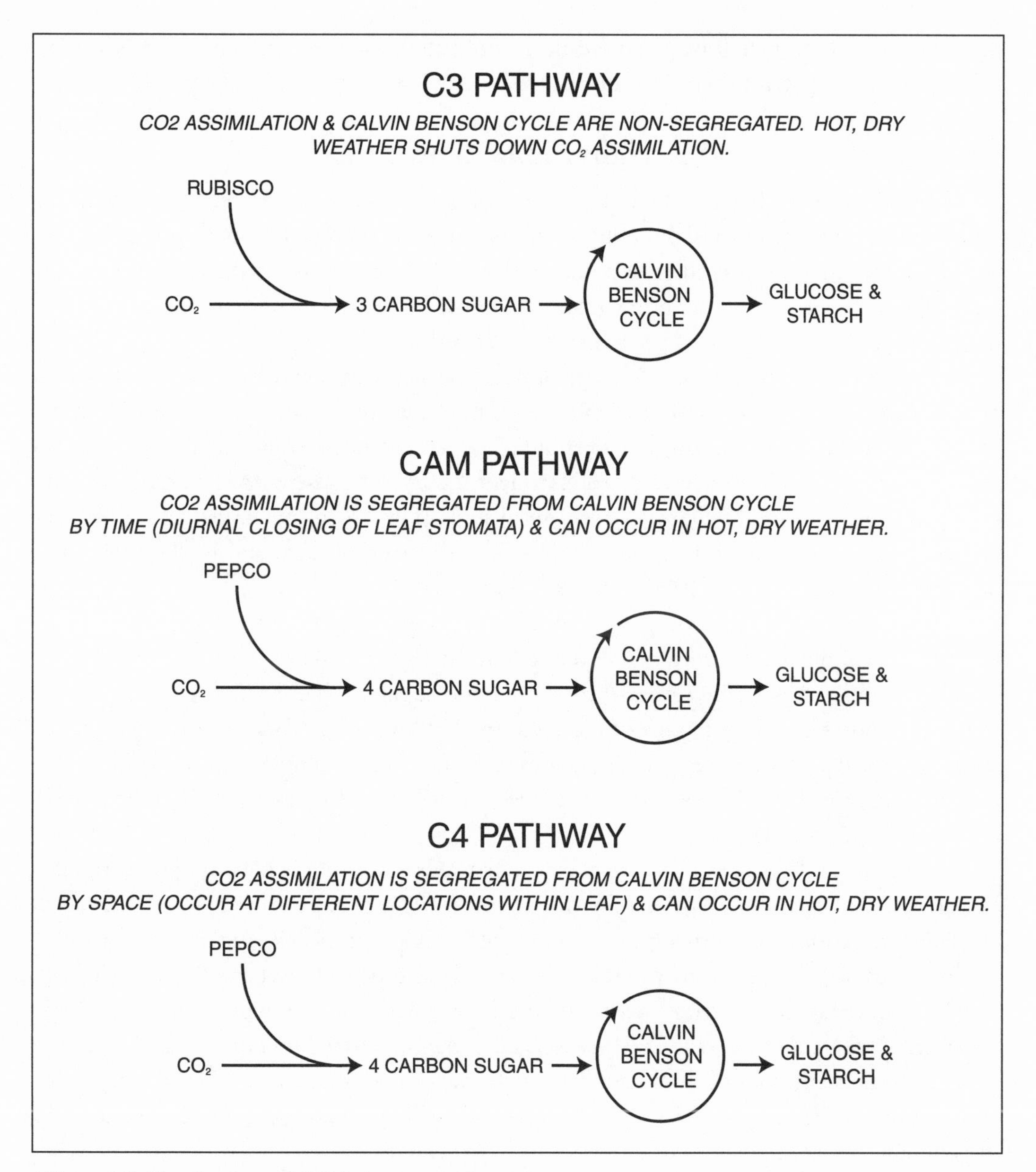

Figure 6.1. *The C_3, C_4, and CAM photosynthetic pathways.*

probability of continuity. Further examination of the genus revealed extensive hybridization among the different species, involving plants with all three types of photosynthesis. This evidence for continuity among the species of *Flaveria* disqualifies C_4 photosynthesis as an indicator of discontinuity.

This example illustrates the necessity of holistic baraminological research to support claims of discontinuity. It is far too easy to equate complexity with created design and to see discontinuity where there is none. Rather than hastily claiming separate creation, baraminologists need to evaluate critically and fairly all of the evidence for and against discontinuity. Only in this way can we be assured that our discontinuity hypotheses are robust and able to stand up to further scrutiny.

6.3. Holism and Consilience

In chapter 2, we defined discontinuity as a holistic difference between two groups of organisms. The holism of discontinuity requires a holistic approach to its identification. As we have just seen in the preceding section, a reductionistic argument for discontinuity (based on a superficial view of one or a few traits) can mislead researchers. Practically speaking, a holistic argument benefits from the **consilience of induction**.

A theory that explains many divergent data is called consilient. Even though the connection between the theory and any individual datum may be tenuous, the correlation of all the data with the theory counts as a separate piece of evidence beyond the directly observable. Evolution, broadly conceived, is a consilient theory. Its consilience is manifested in its ability to explain genetic similarities, anatomical similarities, fossil succession, and ecological **adaptation**, among many other kinds of data. The diversity of data explained under a unified theory testifies to the consilient explanatory power of evolution.

As we have argued previously, an evolutionist views the appearance of discontinuity as imperfection of the fossil record or unevenness in the tempo of divergence. Discontinuity is therefore dismissed as a reflection of our own ignorance rather than reality. As we have seen in the previous section, ignorance of the data can indeed lead to erroneous conclusions of discontinuity. By constructing a theory of discontinuity from multiple lines of evidence, we lend to it a power beyond the individual evidences. It is therefore more likely to be correct and to withstand further scrutiny.

Recognizing the power of a consilient theory, Wise and others have proposed numerous criteria that can be used to identify discontinuity.[3] Using these criteria, baraminologists can construct a consilient discontinuity argument for their organisms of interest (if such discontinuity exists). Enumerating discontinuity criteria also provides a basis for assessing the relative confidence of a discontinuity hypothesis. For example, a discontinuity hypothesis based on 50 percent of the criteria is more reliable than one based on only 20 percent of the criteria. In the remainder of this chapter, we will review and discuss proposed discontinuity criteria in light of the refined baramin concept. We will conclude with a demonstration of discontinuity inference using turtles.

6.4. Discontinuity Criteria

6.4.1. Lack of Continuity

ReMine and Wise proposed that discontinuity[4] may be inferred when there is a "severe failure" to demonstrate continuity. ReMine proposed three continuity criteria: hybridization, known variation, and evidence of lineage.[5] We will discuss hybridization in more detail in the next chapter. The "known variation" criterion requires that potential members of a monobaramin fall within the known variation of previously identified members of that monobaramin. For example, two organisms that share continuity demonstrated by successful hybridization have a number of measurable similarities, including DNA similarity. If a third organism is more similar to one of the hybridizing species than the hybridizing species are to each other, we may conclude continuity even if no hybridization has been demonstrated involving the third species.

Evidence of lineage would include intermediate forms that connect the organism of interest to a known monobaramin. Wood and Cavanaugh recommended generalizing this criterion by removing its phylogenetic interpretation of common ancestry. They called their criterion **biological trajectories**, and it would essentially include any examples of ReMine's lineage criterion.[6]

According to this first discontinuity criterion, the absence of all of these evidences of continuity could constitute evidence of discontinuity. Under the refined baramin concept, discontinuity is defined as a significant, holistic difference, not simply a lack of continuity. As a result, failure to demonstrate continuity would be weak evidence of a true discontinuity. The criterion could be strengthened by demonstrating the failure of all continuity criteria. For example, if hybridization between every member of a group and its most similar outgroup species failed during repeated attempts, this might imply a discontinuity. If tests of continuity fail, other evidences of significant, holistic difference should be sought.

6.4.2. Flood Bottleneck

According to Wise, the population bottleneck induced by the Flood in populations of terrestrial creatures and birds may yield clues to identities of baramins. Specifically, the diversification of organisms after the Flood would probably generate species that are different from the species living before the Flood, even within the same baramin. As a result, members of a baramin found in Flood sediments should be classifiable into a higher taxonomic category with organisms from the post-Flood period. That higher category would therefore approximate a holobaramin, necessarily discontinuous with all other organisms.

Though undoubtedly useful, this criterion might have a limited applicability. It would apply most effectively to terrestrial organisms and birds, baramins that we know for certain underwent a severe bottleneck. Although organisms outside the ark also underwent a bottleneck, we do not know how severe that bottleneck was. Second, it only applies to baramins that have members preserved in Flood sediments. This excludes most mammals and birds, because they are not preserved in Flood sediments in large numbers. Unfortunately, mammals and birds are the very groups we know for certain underwent a severe bottleneck.

Most importantly, we must also recognize that this criterion is an indirect evidence of discontinuity. Identifying a taxon of fossils in Flood and post-Flood sediments does not directly demonstrate significant holistic difference. Like other indirect discontinuity criteria, we can use this criterion as a starting place for statistical studies of similarity, which in turn can reveal significant, holistic dissimilarity.

6.4.3. Biblical

As we have already discussed, the Bible is a useful source of information for the baraminologist. If and when it speaks on a matter, it does so authoritatively. For the biblical criterion, the baraminologist must evaluate evidence from the Scripture for the baraminic status of the group in question. Because many groups are not mentioned in the Bible, the Scripture will often have limited applicability, but the discontinuities that are recorded may be helpful in identifying further discontinuity criteria.

As we have noted in chapters 2 and 3, the broadest groups of organisms created in Genesis 1 form apobaramins because of their separate creation. We may reasonably infer discontinuity between plants, swimming creatures, flying creatures, beasts of the field, and creeping things. Going beyond this may run into problems with interpreting the language as descriptive or classificatory.

In the previous chapter, we outlined methods for obtaining and interpreting biblical data for use in baraminology studies, and we will not reproduce them here. At this point, we encourage baraminologists to err on the conservative side. If the text does not *explicitly* claim discontinuity, do not claim that it does. Remember that the Bible can give useful baraminological information even when it does not specifically claim discontinuity for the group of interest. Broad apobaraminic groups often give excellent starting points for baraminology studies.

6.4.4. Synapomorphies

In cladistic terminology, a **synapomorphy** is a "shared, derived" character and a mark of a monophyletic group. In other words, a synapomorphy is a character that every member of a given group possesses but that closely related outgroups do not. According to evolutionary assumptions, traits like synapomorphies evolve rarely. Thus, if we find a group of species that share a set of traits, they also likely share a common ancestor. Within creationism, synapomorphies could indicate discontinuity if the synapomorphic traits were created by God. Instead of concluding the shared trait signifies a common ancestor, baraminologists conclude that synapomorphies signify a common Creator.

When applying this criterion, we must be careful to avoid essentialist thinking. Essentialists believed that species were defined by a set of invariant characteristics. While it is possible that some characters persisted unchanged since creation, it is more likely that many attributes have changed dramatically. We should also strive for a holistic set of characteristics to infer discontinuity. Just one or two morphological synapomorphies could easily mislead us to infer discontinuity where there is none.

For example, the structures associated with venom injection appear to define the venomous snakes very clearly and distinctly. Because these structures involve the coordination and regulation of numerous genes, cells, and developmental pathways, the venom injection apparatus appears to qualify as a holistic synapomorphy, indicating a discontinuity. Alternatively, it is possible that some members of a baramin might have lost structures. For example, the ability of true flight frequently disappears among island-dwelling insects that belong to baramins that normally fly. If such a complex structural ability as flight can be lost, surely other, more baraminically diagnostic characteristics may also be lost.

In cases where "synapomorphic" traits appear to be lost in some members of a baramin, we can still infer discontinuity by combining evidence of continuity and discontinuity. For example, if most members of a group possess a substantial structural novelty that requires coordination of multiple genes, cells, or developmental pathways, then the structure should be considered an evidence of discontinuity. Because of their complexity, such structures probably arose by God's creation, but may have been lost over time. For organisms that lack this structure, evidence of continuity can clarify their relationship with those organisms that possess the structure. In this way, a holistic discontinuity/continuity argument may be constructed.

6.4.5. Sequence Differences

If discontinuity is holistic, as we have defined it, then it should be readily apparent regardless of the data examined. By comparing the DNA or protein sequences from within a group, we can find a distribution of similarity scores. When we compare the sequences of one group to another discontinuous group, we should find a distribution of sequence similarity scores that is significantly different than the within-group scores. Because this criterion could be "loaded" by comparing the ingroup to *many* outgroup sequences that are very different, we must limit this analysis to the most similar outgroup species.

In a study of mitochondrial DNA from turtles, Robinson discovered sequence differences between turtles and non-turtles.[7] These differences could be observed both in the overall sequence similarity and in more specific analysis of nucleotide substitution patterns. In both cases, not only were turtles most similar to other turtles, but a significant gap existed between turtle/turtle and turtle/non-turtle similarity scores. These results suggest that DNA sequence similarity and difference can be used to identify discontinuities.

6.4.6. Ecology

In 1992, Wise proposed identification of discontinuity from ecological evidence, based on the widely held assumption that the family is a legitimate approximation of the baramin. Because many families exhibit unique ecological preferences, ecological differences might reveal discontinuities. We can also view this criterion as an extension of the biblical criterion. The organisms described in Genesis 1 are divided into groups according to their gross ecological preference (swimming vs. flying vs. creeping). By extension, we may assume that more specific ecological preferences may signal the presence of discontinuity.

Ecological discontinuity may be most apparent for microbial organisms. During the past century, scientists have discovered that bacteria and archaea exploit a wider variety of environments than anyone ever thought possible. Thriving microbial communities have been discovered in hot volcanic springs, at the bottom of oceans, completely engulfed in other cells, even buried in sediments deep underground. Such different environments require holistic differences in the organisms that inhabit them. Consequently, such differences can indicate the presence of discontinuity.

6.4.7. Trophic Level

Based on the same survey of families, Wise proposed that trophic level (e.g., producer vs. consumer) could be used to indicate discontinuity. Like the ecological criterion, this criterion also has limited biblical support. The plants and animals created during Creation Week obviously occupy different trophic categories. As a result, we infer that a difference in trophic level might indicate a separate creation. Additionally, because an organism's trophic category can involve a broad set of characteristics, trophic differences could indicate the presence of a significant, holistic difference.

6.4.8. Ancestral Group Identification

This is the first of three criteria involving study and interpretation of the fossil record. This first criterion involves detection of morphological discontinuity. If the ancestor of the group of interest cannot be identified in the fossil record or even among extant organisms, the group is likely to be discontinuous. We could restate this criterion as simply the existence of a morphological gap separating the group of organisms from all others. This gap ought to be evaluated statistically. Generalizing in this way works very well because evolutionists fail to discover ancestral or transitional forms when a significant morphological gap exists.

6.4.9. Antiquity of Ancestral Group

Like the previous criterion, this one begins with a basic assumption of continuity in cases where an ancestor has been proposed for our group of organisms. If we assume that a series of organisms from the fossil record represents a true phylogenetic series, then we could reasonably expect the fossils to also form a stratigraphic series, beginning with the ancestor in the lowest sediments and ending with the descendent in the highest sediments. We would expect no such pattern from organisms created independently. Evidence of discontinuity may be inferred when the stratigraphically lowest member of the hypothesized ancestral group is higher than the stratigraphically lowest member of the putative descendent. In other words, if all of the fossils of the "ancestral" group occur in the same sediments of the "descendent" group of interest, this could be evidence of independent origin.

Consider the hypothetical example shown in figure 6.2. If B were proposed to be the ancestor of A, we could infer that a discontinuity between A and B exists because the lower end of the stratographic range of B is higher than the lower end of the stratigraphic range of A. In contrast, if groups C, D, and E formed a nice morphological series, the stratigraphy would provide good evidence of continuity. The stratigraphic range of C is lower than that of D, which is lower than that of E. The correlation of morphological similarity with stratigraphic succession provides evidence of continuity, and the lack of such a correlation can indicate a discontinuity. This criterion works best for post-Flood fossils in the Tertiary and Quaternary.

6.4.10. Absence of Stratomorphic Intermediates in Ancestors or Descendants

If scholars believe that a particular group evolved from some ancestral group, the ancestral group should contain fossils intermediate both in form and stratigraphic position. Likewise, the descendent group should also contain fossils intermediate in form and stratigraphy. When these are noticeably absent, baraminologists could interpret the absence as evidence of discontinuity. Since this criterion so closely resembles the previous one, we will elaborate further.

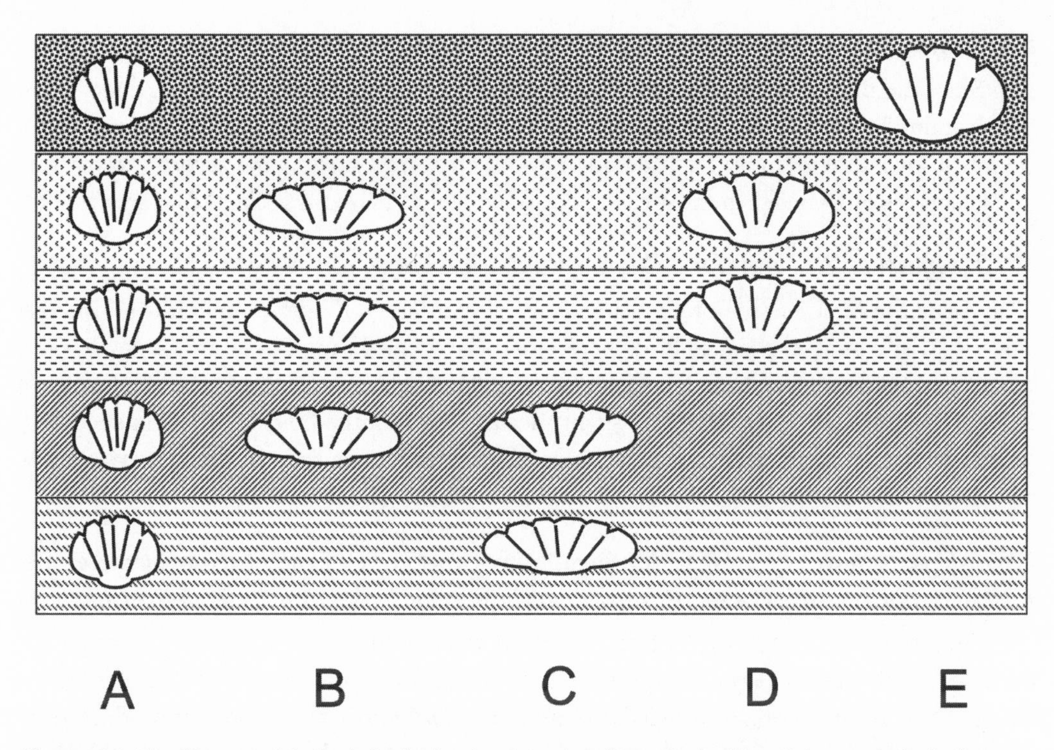

Figure 6.2. The "Ancestral Antiquity" criterion for detecting discontinuity. Taxa B is morphologically thought to be an ancestral group of Taxa A, but Taxa A occurs stratigraphically lower than Taxa B. This could be evidence of a discontinuity between A and B. In contrast, Taxa C, D, and E form both a morphological series and a stratigraphic series, evidence that these taxa belong to the same baramin.

6.4.11. Significant Morphological Difference

As we will explain in chapter 8, baraminologists have introduced several statistical methods for detecting significant similarity and difference between organisms. These methods allow researchers to evaluate morphological similarity for a set of characters from their organisms of interest. When organisms that are potentially discontinuous are included in the analysis, discontinuity can be detected as statistically significant difference in morphology. In chapter 8, we will discuss these methods in more detail.

6.5. The Discontinuity Matrix

To use these criteria most efficiently, it is helpful to organize them into a discontinuity matrix.[8] Each question in the matrix is derived from a discontinuity criterion of the previous section. Answering "yes" to the question is evidence of discontinuity, while answering "no" indicates continuity. The matrix should be used as a guide to discontinuity research, and baraminologists can use copies of the blank matrix as a worksheet to summarize discontinuity evidence. In this section, we will explain the questions and illustrate possible answers from different types of organisms.

6.5.1. Question 1: Does the Bible claim discontinuity for the group?

Since the Bible reveals reliable and infallible information, we begin there. A direct claim of discontinuity would be an explicit act of creation, such as the creation of humans. Also, we might infer discontinuity from the mention of creatures, such as the serpent, in the first three chapters of Genesis. In most cases, the Bible makes no explicit claim to discontinuity.

6.5.2. Question 2: Does the Bible imply discontinuity for the group?

While the Bible might not explicitly claim discontinuity, it might imply it from indirect evidence. For example, because both were present on the ark, the dove and the raven must belong to discontinuous baramins. As we explained in chapter 4, we must take great care not to infer discontinuity when none is present. Because creationists hold the Bible in such high regard, faulty interpretations can be very hard to correct once they have entered the published literature. It is better to err on the side of caution.

6.5.3. Question 3: Do most of the members of the group exhibit a novel metabolic pathway not found in other groups or only in groups known to be discontinuous?

This question is the first of three that derives from the synapomorphy criterion. Because metabolic pathways require the coordination of many genes and proteins, novel metabolic pathways partially fulfill the requirement of holism and probably originated by design only. Consequently, if one group possesses a novel pathway not possessed by any other, that novel pathway might indicate a discontinuity. As we saw in the example of C_4 photosynthesis in *Flaveria,* baraminologists must examine the evidence very carefully to eliminate the possibility of continuity before claiming that discontinuity is present.

6.5.4. Question 4: Is the similarity of ingroup sequence comparisons significantly greater than ingroup vs. outgroup comparisons?

Based directly on the sequence difference criterion, this question will require original research at the website of the National Center for Biotechnology Information (http://www.ncbi.nlm.nih.gov). Users can obtain sequences for a given taxonomic category using the taxonomy browser. Using the BLAST search engine at the same site, we can determine the similarity between a given sequence from a species in our group of interest and other related sequences from a variety of species. We then tabulate these similarity scores in a spreadsheet. Because the distribution of sequence similarity scores is unknown, we recommend using non-parametric statistical tests to determine if the distributions are significantly different.

6.5.5. Question 5: Is the pattern of question 4 true for many genes?

Molecular comparisons generally violate the holism requirement for detecting discontinuity because they focus only on one gene. To overcome this problem, we recommend using multiple genes and repeating the analysis of question 4. If the ingroup comparisons are always significantly more similar than the ingroup/outgroup comparisons, this expands the limited one-gene analysis to a more holistic perspective. Unfortunately, for many groups, multiple gene sequences are not available. Baraminologists will probably encounter sequences from mitochondrial or chloroplast DNA most often. With the recent advances in whole genome sequencing, the lack of data could soon be alleviated even for obscure groups.

6.5.6. Question 6: Do most members of the group possess novel cell types or structures not possessed by other groups or only in groups known to be discontinuous?

With this question, we apply the synapomorphy criterion to the cellular level of biological organization. Are there any organelles or other cell structures that appear to be limited to a particular group? For example, the stinging nematocysts of jellyfish and corals are found only in the phyla Cnidaria and Ctenophora. As such, nematocysts could constitute evidence of discontinuity. We could also apply this to unique cell types or tissues in multicellular organisms.

6.5.7. Question 7: Do most members of the group possess novel organs or anatomical structures not possessed by other groups or only in groups known to be discontinuous?

This question applies the synapomorphy criterion to the anatomy or morphology of organisms. For example, the membranous wings of the bat are novel structures. We could make an argument that they are shared with pterodactyls, but the pterodactyl wing structure is different. We also know from other anatomical evidences that bats and pterodactyls are discontinuous groups. Thus, the bat wing should be considered an anatomical synapomorphy.

Similarly, the structure of the avian lung also marks birds as a discontinuous group. In most vertebrates, the lungs are sacks in which the air is inhaled and exhaled through the same tube (the trachea). In bird lungs, the air passage forms a loop, such that the air enters and exits the lungs through different sets of tubes. In this way, the air passes through bird lungs in a "one-way" fashion. Because this is strikingly different from all other vertebrates, we should consider this as evidence of discontinuity.

6.5.8. Question 8: Is the overall morphological similarity within the group significantly greater than the similarity of the group with other groups?

This question returns to a more holistic perspective by examining the "form" or morphology of the organism, as derived from the criterion of significant morphological difference. To make this question broadly applicable, we could define "form" to include anything that contributes to the outer appearance of the organism, including proteins on the cell membrane as well as visible anatomical structures like arms and legs. Using statistical methods which we will describe in chapter 8, we can examine the overall morphological similarity between many pairs of species. Based on our definition of discontinuity, we would expect that any comparisons within a continuous group would be significantly more similar than comparisons between discontinuous groups. When we find a group that is significantly more similar within the group than it is to members of other groups, this may indicate a discontinuity.

6.5.9. Question 9: Does the group occupy an environment notably different from other organisms?

As noted above, ecology can indicate discontinuity, especially in the case of microorganisms. When comparing our group to "other organisms," we mean other organisms that are not obviously discontinuous. For example, we know from Scripture that plants and animals are discontinuous, so there is no need to look for discontinuity between arctic animals and arctic plants. As in question 4, limit your comparisons to the most similar groups. If the most similar group to your group occupies a different environment, then this may be evidence of discontinuity.

6.5.10. Question 10: Are stratomorphic intermediates that would connect the group to other groups (ancestors or descendants) mostly absent?

This question combines the criteria of ancestor antiquity and absence of stratomorphic intermediates. We reiterate here that we must limit this question to the highest levels of classification that we can. Intermediate species between pairs of genera or other species are comparatively more common in the fossil record than species intermediate between families or orders. Although this sounds like a phylogenetic question, it does fit into our conception of biological character space. Since potentiality regions are by definition bounded by regions where organisms do not exist, the existence of an intermediate organism signals that the apparent gap between two groups might not be a discontinuity. If the group can legitimately be connected with another group, then the two groups may be continuous.

We also ask if the intermediates are "mostly" absent, recognizing that some true stratomorphic intermediates exist between obviously discontinuous groups. The *Archaeopteryx* is both a stratigraphic and a morphological intermediate between birds and reptiles; nevertheless, we know from the biblical record that the birds and creeping things are truly discontinuous. Consequently, the *Archaeopteryx* cannot be accepted as evidence of continuity between birds and reptiles, but it can tell us about the nature of discontinuity and its detection.

We note from this example that comparatively few intermediates between major groups have been discovered. Surely if some group of reptiles evolved into birds, a large number of intermediates must have been necessary to bridge the large morphological gap. Because so few have been discovered, we should hardly consider it a baraminic intermediate. We might formalize this observation by requiring that continuity be demonstrated by discovery of a number of fossils that is proportional to the size of the morphological gap. Since the number of intermediates (e.g., *Archaeopteryx*) is not proportional to the wide gap between reptiles and birds, we argue that it does not constitute evidence of continuity.

6.5.11. Question 11: Is the lowest member of the proposed ancestral group found in a higher layer than the lowest member of the group of interest?

We adopt this question virtually unchanged from the discontinuity criterion with one precaution. As with the previous question, we require that it be applied at a high taxonomic level. Rather than look for interspecific intermediates, we encourage baraminologists to scan the literature for interfamilial or interordinal intermediates. The absence of these higher level intermediates more likely reflects true discontinuity than missing interspecific intermediates.

6.6. Using the Matrix

Before we use the matrix, we must understand that the questions are not all equally important. Evidence gathered for some questions can be more important than for others. Obviously, when the Bible clearly claims discontinuity, any other evidence is unnecessary. As a result, the quality of the *Australopithecus* or whale stratomorphic series is overruled by the biblical claims of discontinuity between humans and apes and between whales and land creatures. By extension, it is likely that some of the questions in the discontinuity matrix will not be as important as others. Only with use will baraminologists be able to learn which questions are the most useful.

We also must view the matrix only as a guide to practical discontinuity detection. New questions could be added, and some questions could be eliminated. The matrix is not a fail-safe test for discontinuity but a tool to help baraminologists retain that all-important holistic perspective. As we noted in previous sections, holistic, consilient arguments are more powerful and explanatory than their reductive counterparts. Claims of discontinuity based on single characters are often easy to discredit. By approaching the problem holistically, as the matrix helps us to do, we are more likely to produce a robust theory of discontinuity.

Practically speaking, discontinuity detection may be accomplished by using the matrix as a guide. If we can answer more than one question in the affirmative, it may indicate discontinuity is present. The strength of our discontinuity hypothesis should be proportional to the number of questions that can be answered positively. As we noted in the previous section, we can also compare hypotheses of discontinuity. Suppose we find discontinuity surrounding a group based on eight of the eleven questions. If subgroupings of the same group have only four or five affirmatives out of eleven, then we would conclude that the evidence for discontinuity between the subgroupings is less reliable than the evidence for discontinuity of the group itself. We must be careful in drawing these types of conclusions because some questions are more important than others. We cannot say that discontinuity supported by eight of eleven questions has 8/11 = 73 percent support, but we can say that such discontinuity is well-supported. Once again, the matrix is only a guide, not a test.

6.6.1 Applying the Matrix to the Turtles

To conclude this chapter, we will briefly illustrate the use of this new discontinuity matrix. Following Wise's original example, we will use the turtles, order Testudines, as our test subject. Wise chose the turtles because of their historical importance as a subject of research by Wayne Frair, as we noted in chapter 6. For questions that cannot be answered with a simple "yes" or "no," we mark them as "N.A." (Not Applicable). For questions that cannot be answered with present data, we mark them "U" (Unknown).

6.6.1.1. Biblical discontinuity? The turtles are not mentioned in the Bible, other than the generic term "creeping things" (*remes*). Because the Bible does not indicate discontinuity between lizards and turtles, we conclude that the Bible neither claims nor implies a discontinuity surrounding the turtles. As a result, we mark questions 1 and 2 as "Not Applicable."

6.6.1.2. Novel metabolism? Although we know of no metabolic synapomorphies in the turtles, such novelties might exist. Consequently, we mark question 3 as "Unknown."

6.6.1.3. Sequence difference? Based on DNA/DNA hybridization and blood serum reactivity experiments, we may conclude provisionally that turtles and non-turtles are significantly different at the molecular level.[9] Furthermore, Ashley Robinson's analysis of turtle mitochondrial DNA provided further support for a discontinuity in sequence similarity.[10] We therefore mark question 4 with a "yes." Question 5 asks whether the discontinuity observed in question 4 holds true for many genes. Because DNA/DNA hybridization approximates similarity across entire genomes, we can also mark question 5 with a "yes."

6.6.1.4. Novel cell types or structures? As with metabolism, we are unaware of any cellular or subcellular synapomorphies that define the turtles. We must mark question 6 as "Unknown."

6.6.1.5. Novel anatomical features? Unlike all other vertebrates, the turtle pectoral and pelvic girdles are on the inside of the rib cage, which is modified as part of the turtle's shell. The shell is also partially composed of dermal bone.[11] As a result, we may confidently mark question 7 as "yes."

6.6.1.6. Significant morphological difference? Although no baraminology studies have been performed on turtles using the new statistical methods, we can intuitively answer question 8 affirmatively. The presence of the shell and the limb girdles inside the rib cage for all turtles means that morphological similarity among the turtles will certainly be higher than any turtle/non-turtle similarity.

6.6.1.7. Ecological discontinuity? Wise noted that turtles are potentially ecologically discontinuous with other organisms.[12] We will provisionally mark question 9 with a "yes."

6.6.1.8. Fossil discontinuities? Based on Wise's summary of turtle fossil evidence, stratomorphic intermediates connecting turtles with non-turtles are rare. In particular, the Triassic *Proganochelys* is acknowledged as the first turtle, with a complete turtle shell. The ancestors of *Proganochelys* are both uncertain and not clearly linked by stratomorphic intermediates.[13] From this we conclude that stratomorphic intermediates do not connect turtles with putative ancestors, and we mark question 10 with a "yes." Despite uncertainty of the relationship of turtles to other organisms, all proposed ancestors of the turtles are in lower stratigraphic units than *Proganochelys*. Consequently, we mark question 11 with a "no."

6.6.1.9. Summary. In this brief review of turtle discontinuity, we have six affirmative answers out of nine applicable criteria in our completed discontinuity matrix for turtles. This result strongly indicates discontinuity between turtles and other organisms. Unlike other examples of discontinuity illustrated in previous sections, this conclusion can be easily defended since it is supported by multiple types of data: molecular, morphological, and fossil.

6.7. Chapter Summary

Discontinuity is defined as "a significant difference between two sets of organisms detected in a holistic analysis." The baraminologist's acknowledgment of true discontinuity contrasts starkly with the universal common ancestry of evolution. Unfortunately, it is all too easy to assume discontinuity where there is none. Although God's biological

Discontinuity Matrix

	Y	N
Question 1: Does the Bible claim discontinuity for the group?		
Question 2: Does the Bible imply discontinuity for the group?		
Question 3: Do most of the members of the group exhibit a novel metabolic pathway not found in other groups or only in groups known to be discontinuous?		
Question 4. Is the similarity of ingroup sequence comparisons significantly greater than ingroup vs. outgroup comparisons?		
Question 5: Is the pattern of question 4 true for many genes?		
Question 6: Do most members of the group possess novel cell types or structures not possessed by other groups or only in groups known to be discontinuous?		
Question 7: Do most members of the group possess novel organs or anatomical structures not possessed by other groups or only in groups known to be discontinuous?		
Question 8: Is the overall morphological similarity within the group significantly greater than the similarity of the group with other groups?		
Question 9: Does the group occupy an environment notably different from other organisms?		
Question 10: Are stratomorphic intermediates that would connect the group to other groups (ancestors or descendents) mostly absent?		
Question 11: Is the lowest member of the proposed ancestral group found in a higher layer than the lowest member of the group of interest?		

design displays a staggering complexity, sometimes continuity can bridge incredible "gaps" in biological character space. The holism of discontinuity requires a holistic approach to its identification.

The Discontinuity Matrix is a convenient tool to organize more efficiently the criteria for discontinuity that have been proposed by creationists over the years. It should be used as a guide to discontinuity research, and baraminologists can use copies of the blank matrix as a worksheet to summarize discontinuity evidence. The matrix is composed of a series of questions:

Question 1: Does the Bible claim discontinuity for the group?

Question 2: Does the Bible imply discontinuity for the group?

Question 3: Do most of the members of the group exhibit a novel metabolic pathway not found in other groups or only in groups known to be discontinuous?

Question 4: Is the similarity of ingroup sequence comparisons significantly greater than ingroup vs. outgroup comparisons?

Question 5: Is the pattern of question 4 true for many genes?

Question 6: Do most members of the group possess novel cell types or structures not possessed by other groups or only in groups known to be discontinuous?

Question 7: Do most members of the group possess novel organs or anatomical structures not possessed by other groups or only in groups known to be discontinuous?

Question 8: Is the overall morphological similarity within the group significantly greater than the similarity of the group with other groups?

Question 9: Does the group occupy an environment notably different from other organisms?

Question 10: Are stratomorphic intermediates that would connect the group to other groups (ancestors or descendants) mostly absent?

Question 11: Is the lowest member of the proposed ancestral group found in a higher layer than the lowest member of the group of interest?

Review Questions

1. What is the most common mistake among creationists inferring discontinuity?
2. Why is it dangerous to define a baraminic group with only a single characteristic?
3. What is consilience of induction and why is it important?
4. How can the Flood bottleneck aid in identifying holobaramins?
5. What did Wise discover by studying the ecological preferences of families?
6. What did Wise discover by studying the trophic level of families?
7. What can we infer about discontinuity from the Creation account of Genesis 1?
8. What is a biological trajectory?
9. What is a synapomorphy? How can they be used to infer discontinuity?
10. List three continuity criteria discussed in this chapter.
11. Give an essentialist explanation of species.
12. Does essentialism adequately describe species?

13. How can analysis of DNA establish discontinuity?
14. Where is ecological discontinuity most obvious?
15. How does stratigraphy help determine continuity or discontinuity?
16. How does ecology indicate discontinuity?
17. How is the strength of a discontinuity hypothesis related to the number of positive affirmations within the matrix?
18. Why can the example of discontinuity in turtles be so easily defended?

For Further Discussion

1. Give a biblical justification for the ecological criterion.
2. What criteria might you use to construct a Continuity Matrix?
3. Justify the study of the pattern of life by God's desire to be known.
4. Propose new discontinuity criteria and develop a revised Discontinuity Matrix.
5. Explain and justify ways in which discontinuity could be inferred from Scripture.
6. Explain how the Discontinuity Matrix helps to prevent hasty inferences of discontinuity. Explain how it could be misused to infer discontinuity within a continuous group.
7. Explain how our biases and expectations might inappropriately guide our discontinuity research.
8. How do seals, sea lions, and walruses fit into the Genesis creation account? Are they land dwellers or ocean dwellers? Justify your answers. What might these organisms indicate about the fossils believed to be evolutionary transitions between land mammals and whales?
9. How can we effectively use the matrix if all the questions do not hold equal importance? How can we avoid inserting our own biases into the analysis?

CHAPTER 7
Hybridization

7.1. Introduction

Creationists have not been the only proponents of utilizing hybridization information in systematics, but creationists have stood alone in their fervor to use hybridization as the foundation of systematics. Due largely to Marsh's enthusiasm, hybridization has enjoyed a very exclusive claim on creationist systematics for decades. Although hybridization remains one of the most often used techniques in creationist systematics, the introduction of baraminology reduced its importance to merely one way to identify monobaramins. The interest of baraminologists shifted away from hybridization due in part to theoretical problems and in part to practical limitations. The problems were known to Marsh, but creationists were slow to address these limitations.

7.2. Hybridization in Systematics

Until recently, many non-creationist scientists have not recognized interspecific hybridization as the extensive biological phenomenon that it is. For example, many general biology textbooks continue to present hybridization only in the context of the reproductive isolation of species. To reinforce the reality of species and the legitimacy of the biological species concept, textbooks often minimize their discussion of hybridization, giving the false impression that it occurs only rarely. In reality, hybridization occurs with surprising frequency, particularly among plants. Botanists estimate that up to 70 percent of angiosperm species arose through hybridization of two other species, some of which could also be of hybrid origin.[1] Even among animals, many species readily hybridize. Annie Gray compiled lists of more than five hundred hybrids of mammals and fifteen hundred bird hybrids in the 1950s.[2]

Studies of hybrids often focus on speciation and reproductive isolation, revealing numerous important trends. For example, hybridization often results in sterile offspring. The mule, a cross between the horse and the donkey, usually does not bear offspring. Likewise, the hybrid of the radish and cabbage is viable but sterile. In both of these cases, the chromosomes differ to the extent of halting meiosis, resulting in a lack of gametes. In cases where hybridization produces fertile offspring, a number of different types of scenarios can arise. When hybridization occurs often in a limited contact zone between the two parent species, a sterile population of hybrids, known as a **hybrid swarm**, can result. In some cases, hybrids can actually be heartier and more fertile than either parent species, a condition called **hybrid vigor** or **heterosis**.

Many other hybrids are reproductively fertile. When hybrids cross with members of either parent species, **introgression** may take place. Introgression is the permanent introduction of genes from one species into the genome of another. Introgression may lead to one of two outcomes. First, the parental species could eventually merge into one species if no isolation mechanism keeps them apart. Second, the introgression could be temporary, followed by a reinforcement of the reproductive isolation of the parent species. In such cases, the introgressed genes merely enrich the genetic diversity of the species involved.

Fertile hybrids might also form new species, especially in cases of hybrid vigor. Forming of hybrid species is especially common in plants, which can undergo chromosome doubling. For example, the sterile hybrid of the radish and cabbage can become a fertile species (genus *Raphanobrassica*) if the chromosomes double. In other cases, the fertile hybrid could be uniquely adapted to life in a different habitat, allowing it to exploit a niche with little competition from either parent.

Some scientists have recognized the commonplace nature of hybridization and have attempted to incorporate interspecific crossing into formal systematic definitions. Within the

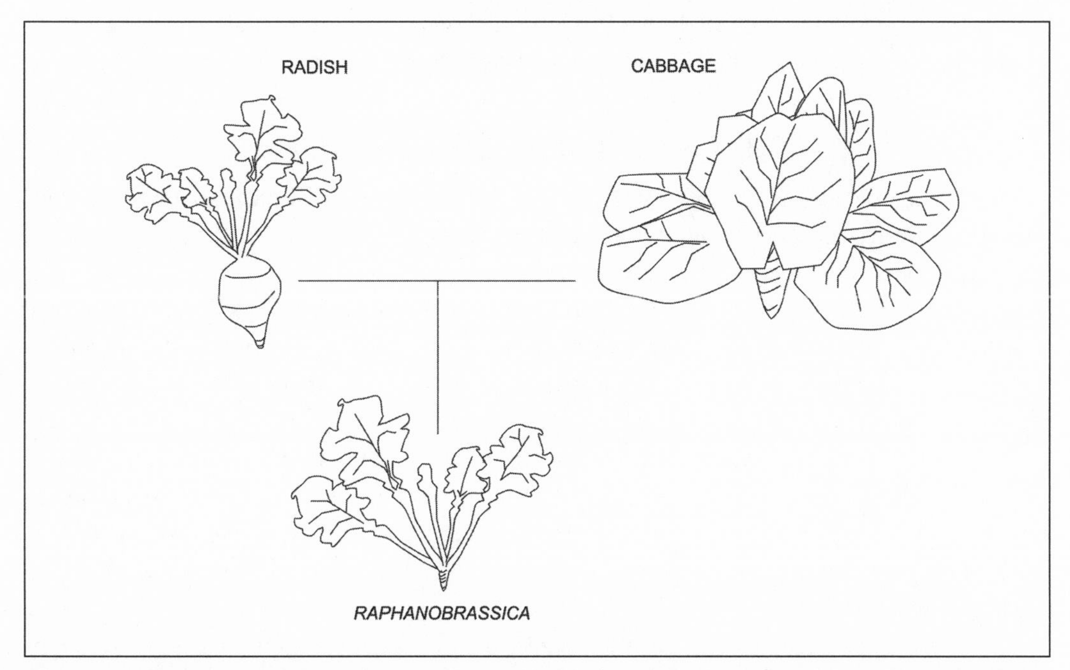

Figure 7.1. *Though morphologically dissimilar, the radish (genus* Raphanus*) and the cabbage (genus* Brassica*) are able to cross and produce fertile hybrids, indicating that the parent species belong to the same monobaramin.*

evolution model, as two species diverge from a common ancestor, their genes and chromosomes also change. In the days before genome sequencing, hybridization was used to gauge the degree of genetic and chromosomal diversification. Growth and development require compatible genetic material from both parents, and gamete production requires that the chromosomes be similar enough to pair during meiosis. Thus, viability and fertility of hybrids reflects the degree of genetic and chromosomal differences between the parent species. For example, species that produce fertile hybrids must have very similar genetic material and are probably closely related. Species that produce nonviable or sterile hybrids are more distantly related.

In the 1930s, Clausen, Keck, and Hiesey performed systematic hybridization experiments in the subtribe Madiinae of the sunflower family.[3] Their goal was to experimentally define taxonomic groups. They called species that produce fertile hybrids *ecotypes* or *biotypes,* which they equated with the subspecies concept of traditional taxonomy. They proposed ecospecies to describe species that could produce partially sterile offspring and cenospecies to refer to species that produced only sterile hybrids or none at all. To differentiate cenospecies that could produce sterile hybrids from those that could not, they coined the term **comparium**. Two cenospecies capable of hybridizing belong to the same comparium. Cenospecies that could not hybridize belonged to separate comparia.

The comparium concept never became widely used, but later researchers also recognized the value of hybridization as an indicator of systematic relationship. Grant proposed the term syngameon to describe a group of hybridizing species, and some plant systematists have adopted and used this term.[4] Scherer cited several specialists who argued for the use of hybridization in recognition of genera.[5] In all of these examples, researchers use hybridization as a means of gauging the diversity of chromosomes. Creationists alone have assigned comparia or syngameons with the deeper significance of a created unit.

7.3. Re-envisioning Hybridization

Despite practical difficulties that we will discuss below, hybridization provides very powerful additive evidence when it is available and when it works. According to the refined baramin concept, the species of baramins and monobaramins share significant, holistic similarity (continuity). The successful production of hybrid offspring provides evidence of continuity because it involves the coordination of many levels of organization in the organism. Production of hybrids depends on the cooperation of genes, developmental pathways, and ultimately the form or morphology of the adult organism.

Without hybridization, baraminologists would encounter the more formidable task of demonstrating significant similarity between the genetics and development of two different species. Hybridization offers a practical method of doing just that. As a result, we encourage baraminologists to use hybridization wherever possible, as long as its limitations are recognized and respected.

The high frequency of hybridization leads us to an important question that we ought to consider: Why do we have species at all? This question should puzzle creationists more than evolutionists because of the limited time we have available to produce modern species diversity within each baramin. For an evolutionist, species have millions of years to develop and maintain reproductive barriers. For creationists, the diversification of baramins into stable species occurred with a startling swiftness. If we only rely on chance or neodarwinian speciation mechanisms, how could species develop so quickly and still retain the ability to cross successfully with related species?

Although we will provide a more comprehensive answer to this question in chapter 11, we may infer some important observations that will aid our understanding of this peculiar problem. Most obviously, since interspecific hybridization can occur, sometimes with ease, species must not primarily be reproductive units. Certainly reproductive isolation contributes to the present stabilization of species phenotypes, but reproductive isolation could not be the originator of species.

According to the evolutionary model of speciation, reproductive isolation is the primary cause of speciation because species are defined as reproductively isolated groups. Stable phenotypic diversity arises only after the origin of reproductive isolation. Because of the limited time available for speciation in the creation model, reproductive isolation is more likely the result of phenotypic diversification, rather than its cause. If diversity of form precedes reproductive isolation, then we would expect hybridization frequency to be high among members of the same baramin.

7.4. *Practical Problems with Hybridization*

As we noted above, the limitations presented here do not undermine the use of hybridization when available, but merely limit its applicability within biology. As we use hybridization information, we must remember these limitations and strive for a holistic definition of the baramin. We certainly should not rely on hybridization as the exclusive method of baraminology. Instead, we should use a wide variety of evidence in addition to hybridization to define holobaramins and baramins.

7.4.1. Hybridization only applies to living, sexual organisms

Most obviously, baraminologists can only use hybridization information for organisms that can hybridize. This limitation necessarily eliminates strictly asexual organisms like bacteria and archaea, predominantly asexual organisms like many protists (and even a few fungi, plants, and animals), and fossils of species that are presently extinct. In summary, hybridization can be used on lots of animals and plants, some algae and fungi, and a few protists. While these groups represent probably hundreds, or even thousands, of baramins, members of many more baramins simply cannot or do not cross.

7.4.2. Hybridization may not be practical or justifiable

Operational difficulties can often make hybridization experiments difficult or nearly impossible. Even among the sexual organisms that have the ability to create hybrids, technical difficulties often prevent meaningful progress in hybridization experiments. This problem most often arises in dioecious or hermaphroditic organisms capable of self-fertilization and in organisms from which mature, viable gametes are difficult to isolate and sustain in the laboratory. In both such cases, no biological problem prevents the gathering of hybridization information, but the techniques involved in obtaining hybrids can pose significant barriers.

Likewise, the reproduction of many endangered species is closely guarded by conservation advocacy groups. In those cases, baraminologists will find substantial opposition to obtaining gametes for the purpose of hybridization experiments. For example, the baraminic position of the giant panda will most likely be resolved by means other than hybridization merely because the species is endangered.

7.4.3. Hybridization failure is ambiguous

Many factors unrelated to baraminic status can cause hybridization to fail even in experimental settings. For example, salamanders of the family Plethodontidae perform complicated mating dances prior to fertilization.[6] Female plethodontids respond poorly to the dances of males of different species, and males will often not dance for females of different species. Consequently, fertilization does not occur, and hybridization records are extremely rare for these salamanders. In other cases, differences in the proteins on the surface of the sperm or egg can prevent hybridization. Sometimes this problem can be overcome with persistence on the part of the researcher performing artificial fertilization. Chromosomal differences can often disrupt the chromosome pairing during mitosis, rendering hybridization improbable even with experimental persistence.

Because numerous factors influence hybridization success, we cannot consider hybridization failure alone as evidence of anything. As we have noted all along, however, a holistic observation of lack of hybridization success might be considered weak evidence of discontinuity. For example, if we could test cross every member of a particular monobaramin with individuals from the most similar outgroup species, failure to hybridize in *every* case might constitute limited evidence of discontinuity. We should not be dogmatic about such an observation because the factors discussed above can render a species incapable of crossing even with other members of the same baramin.

7.4.4. Hybridization success is difficult to define

Last, and perhaps most importantly, the definition of hybridization success is difficult to define. Most baraminologists agree that a sterile hybrid constitutes a success, and many would admit that a naturally aborted fetus also represents a successful cross. Going to earlier stages in embryonic development raises many questions. Marsh believed that true fertilization constituted evidence of hybridization, but numerous modern observations seem to disqualify this idea. Most importantly, we now know that human sperm can fertilize hamster ova, though no development occurs.[7] Since we know that God created humans separately from hamsters, the "true fertilization" must not be admissible as evidence of common baraminic membership. More recently, Scherer proposed that the developing embryo must reach a stage in which the coordinate expression of genes from both parents must be detected.[8] This may be late in development, since frog development can proceed through twelve cell divisions even when the nucleus is completely absent.[9] A proper definition of "successful hybridization" remains elusive.

7.5. Locating Hybridization Information

Many biologists have a keen interest in hybridization. The emphasis on gene flow in modern species concepts inspires many ecologists to study naturally occurring hybrid zones. Systematists study natural hybrids to identify species. Genomics researchers create hybrids in order to develop genetic maps. Animal and plant breeders hybridize species to generate hardier breeds of crops and livestock. As a result of so much research in hybridization, the published technical literature contains thousands of records of successful hybrids. This sounds like good news to baraminologists, but locating hybridization records can be a daunting task. In this section, we will offer advice on how to find records of hybridization.

7.5.1. Online Sources

The HybriDatabase (HDB) is the premiere source for hybridization information, specifically designed by and for baraminologists (http://www.bryancore.org/hdb/).[10] Striving to become a comprehensive source of information, the HDB integrates hybridization records from all disciplines, including genomics, ecology, and plant and animal breeding. Currently, the HDB contains over five thousand hybrid records, including many plants, animals, and birds. The HDB curation team at the Center for Origins Research and Education at Bryan College has many more records in their files that they will include in regular updates to the HDB.

For each hybrid cross, the HDB contains information that baraminologists can immediately add to their research projects. Both parent species are listed, and if available, the database also contains information on the sex of the parents, the success of the cross, whether the cross was artificial or natural, and the reliability of the report. In addition to crosses that produced offspring, the database also contains crosses that failed to produce any offspring; thus, baraminologists must carefully examine each listed cross. Every cross in the database links to a complete literature citation for further research. Baraminologists who choose to use the HDB ought to double-check each hybrid reference in the original literature. Although the curation team makes every effort to include accurate information, sometimes errors occur.

Using the HDB, baraminologists can search for a genus or an author. When users select "genus," the database searches all genera listed in the HDB. "Author" searches examine all references contained in the HDB. Both searches return a complete list of recorded hybrid crosses in tabular format, with the total number of records retrieved indicated at the top of the output page. A detailed tutorial on using the HDB can be found at the HDB website, along with project information and updates from the curation team. The site also contains a complete list of project personnel who have contributed to the development of the current HDB.

Because the HDB project has only just begun, many thousands of records remain to be entered. To identify other hybrids not listed, several other internet-searching techniques can be employed. Baraminologists should first consult databases of published literature. The National Center for Biotechnology Information offers free searching of thousands of abstracts related to biomedical research in their PubMed database (http://www.ncbi.nlm.nih.gov/pubmed/).

Because hybridization can aid in development of genetic maps, PubMed contains occasional references to interspecific hybrids. Baraminologists should be wary and careful using PubMed, because molecular biologists also use the term *hybridization* to refer to the process of base-pairing between two complementary strands of DNA or RNA. As a result, searching for just the term *hybridization* can yield hundreds of positive matches. Baraminologists should qualify their literature searches with the words *intergeneric* or *interspecific*. In all web searching, include the name of the genus or family of interest. For example, baraminologists interested in herpetology might search for "naja interspecific hybridization" to locate references to hybrid cobras (genus *Naja*).

For more general searches of the biological literature, options available without a fee are limited. Several newer websites offer searching of a limited selection of general biology literature. Ingenta offers searching of a database of ten million articles from 1988 to the present, covering many technical disciplines (http://www.ingenta.com). For baraminologists looking specifically for literature on reptiles or amphibians, consult the HerpLit database (http://www.herplit.com), with fifty thousand references from the sixteenth century onward. As with PubMed, baraminologists should remember to search for intergeneric or interspecific hybridization.

For more comprehensive searches of the biological literature, consult *Biological Abstracts,* currently partially available as the Biosis Biological Abstracts Database (http://www.biosis.org). The online database, available only through a subscription fee, contains approximately six million articles back to 1980. The printed version of *Biological Abstracts* began publication in 1927 with the merger of *Abstracts of Bacteriology* (1917–1925) and *Botanical Abstracts* (1918–1926). Although the online version has a limited coverage, the print version contains a wealth of older information that can be obtained no other way. Most public university libraries subscribe to the online version of *Biological Abstracts* as well as maintain the back issues of the printed version.

A few other resources warrant mention. The Current Contents database, available for a subscription fee from the Institute for Scientific Information, provides online searching of many science journals published in the last year. Public university online library catalogues can also be a useful source of references to hybridization. Finally, baraminologists also may consider searching for hybrid records using the booksellers' resource Bookfinder (http://www.bookfinder.com). Bookfinder provides a centralized, searchable inventory of approximately forty thousand book dealers from around the world. Information about each book available is limited, because the dealers often enter only the title, author, and date. If a title sounds particularly useful, users can contact the dealer directly for more information.

While searching for hybridization records online, the baraminologist should also consult one or more of the many general internet search engines, such as Google (http://www.google.com), Yahoo (http://www.yahoo.com), or AltaVista (http://www.altavista.com). Scientific societies and individual scientists often post bibliographies or personal reference lists on the internet, sometimes as a service to their members (for societies) or as part of their resumes (individuals). A few societies publish newsletters and include full-text content online. Since newsletters are rarely indexed in any database, web searching may be the only way to locate these references. By searching for the same combination of terms used in any search of the specialty databases, baraminologists might find obscure references to unusual crosses.

7.5.2. Printed Resources

We have already mentioned one printed resource, *Biological Abstracts,* but there are a number of other resources that provide especially useful references for the baraminologist. Two books published by Annie Gray in the 1950s cover hybrid mammals and birds. *Mammalian Hybrids* (1954, revised 1972) contains approximately five hundred hybrid records from a variety of mammal groups. Although all crosses recorded in *Mammalian Hybrids* are also listed now in the HDB, baraminologists can consult the original book for additional literature references to any hybrid of interest. *Bird Hybrids* (1958) contains approximately fifteen hundred records, which are also contained in the HDB. Like *Mammalian Hybrids, Bird Hybrids* also contains extensive references not currently listed in the HDB.

For information on plant hybrids, the numerous horticultural societies often maintain registries of the many cultivated hybrids. For example, the Royal Horticultural Society (http://www.rhs.org.uk) keeps hybrid registers for *Clematis,* conifers, dahlias, larkspurs, *Dianthus,* lilies, daffodils, orchids, and rhododendrons. Each of these registries may be purchased in printed format or may be searched online. For hybrids of the grass family Poaceae, consult Irving Knobloch's 1968 volume *A Check List of Crosses in the Gramineae.*[11] Because of the popularity of horticulture as a hobby, many other hybrid checklists and compilations like these may be available to borrow from local libraries or to purchase from one of the dealers in the Bookfinder inventory.

Since many larger zoos often sustain an active breeding program, the *International Zoo Yearbook* can be a source of useful information about crosses of more obscure animals. For example, the 1969 yearbook included a reference to an unusual intergeneric hybrid involving a male polar bear (*Thalarctos maritimus*) and a female brown bear (*Ursus arctos*). The brief report lists thirteen offspring born over a period of five years at the Lodz Zoo in Poland.[12] Any of the other sources already discussed should turn up a plethora of references to crosses involving any kind of livestock, but for exotic creatures, unusual hybrids often appear in the *Zoo Yearbook,* published from 1959 to the present.

For a comprehensive source of information on hybridizations, consult the *Plant Breeding Abstracts* (PBA) or *Animal Breeding Abstracts* (ABA), published by CAB International Publishing (http://www.cab-publishing.org). PBA and ABA both publish full abstracts in English from hundreds of international journals, focusing primarily on the subject of breeding. Recent copies of these periodicals contain hundreds of references to interspecific and intergenetic hybridization. PBA began publication in 1930, with ABA beginning in 1933. For each publication, CAB International also provides an electronic version with a ten-year archive. Many public university libraries maintain subscriptions to these publications.

In addition to these general resources, many libraries have bibliographies or abstract collections that pertain to specific groups of organisms. As we mentioned previously, searching through a library catalogue can sometimes yield useful information. In all cases of hybrid reports, scan the references of the report for other references. Additionally, papers on organismal systematics often contain references to hybridization. Following "literature trails" can sometimes be more rewarding than broader and less time-consuming database searches. By following every lead available, the diligent baraminologist should be able to create a fairly comprehensive list of known hybrids for any group of interest. To benefit fellow baraminologists, contribute hybrid lists to the HybriDatabase.

7.6. Using Hybridization Information

Before we can discuss how to best utilize hybridization, we must decide what constitutes a successful hybridization. Hybridization is a process that begins at fertilization and ends at birth. At any stage during this process, the developing organism could die, and hybridization would fail. As we mentioned above, those who rely on hybridization exclusively to define baramins must find a meaningful definition of hybridization. For baraminologists, the problem is less critical (because we can use other criteria) but still important. Here, we do not intend to propose an all-inclusive definition of hybridization success, but we do want to present a few principles that can guide the interpretation of hybridization.

Most hybrids reported in the literature are live-born individuals, sometimes sterile but often fertile. In these cases, the hybridization clearly succeeded. Some citations report on naturally aborted hybrid embryos; others discuss **apomixis**. Apomixis occurs when the chromosomes of one of the parents are eliminated from the cells, giving rise to a haploid individual. We can easily see that apomixis is not hybridization at all, but aborted embryos pose a potentially more difficult theoretical challenge. To address this problem, we must return to our refined baramin concept.

Because we define continuity as significant, holistic similarity, we would recommend that development progress to an advanced stage before considering it a success. For example, Marsh accepted only "true fertilization," and Scherer qualified that by requiring that genes from both parents be coordinately expressed. Both of these cases are less holistic than a hybrid individual brought to term, in which the full developmental mode has been expressed. In the end, we may simply define hybridization success relatively rather than absolutely. In other words, instead of arbitrarily choosing a developmental stage to be the definition of hybrid success, strive for a holistic baraminology argument. If the hybridization potential between two species or groups is dubious, try to develop an argument for baraminic relationship from other additive methods. Above all, remember that the baraminologist has a whole palette of baraminology methods and tools at his disposal. The success of baraminology does not rise and fall with the success of hybridization.

7.6.1. The Hybridogram

Having selected "successful" hybrids to use as additive evidence, the baraminologist must then display the data in order to infer monobaraminic groupings. Most baraminologists employ a graphical device that has come to be called a **hybridogram** (also called a "cross-breeding matrix"). Originally employed by conventional biologists,[13] the hybridogram has become a very useful tool in baraminology studies. A hybridogram is a simple matrix, with species listed down the side and across the top. Successful hybridizations are indicated with a black box at the intersection of the row and column corresponding to the hybrid's parent species. In the case of very large baramins, hybridograms could also be constructed by using a higher taxonomic category (such as genus) rather than species.

Hybridograms are most easily constructed using any modern spreadsheet software, such as Corel QuattroPro or Microsoft Excel. Using the cell shading function, successful hybrids can be recorded with a click of the mouse. With a portable notebook computer, baraminologists can list their species and update the hybridogram as they discover more reports in the primary literature. To convert an Excel spreadsheet to a graphical image, use an image editing program such as Adobe Photoshop, Corel Draw, or even the Paint program that comes with most versions of MS Windows. Simply highlight the hybridogram in Excel, copy the information to the clipboard, then paste into a new image in the image editing software. MS Windows will automatically convert the information into a bitmap image format.[14]

The hybridogram for the finch family Fringillidae illustrates the utility of the hybridogram (Figure 7.2).[15] The diagram draws the viewer's eye immediately to the successful crossing of the canary (*Serinus canaria*) with nearly every other species in the matrix, including members of both subfamilies. From the hybridogram, we also see that recorded canary

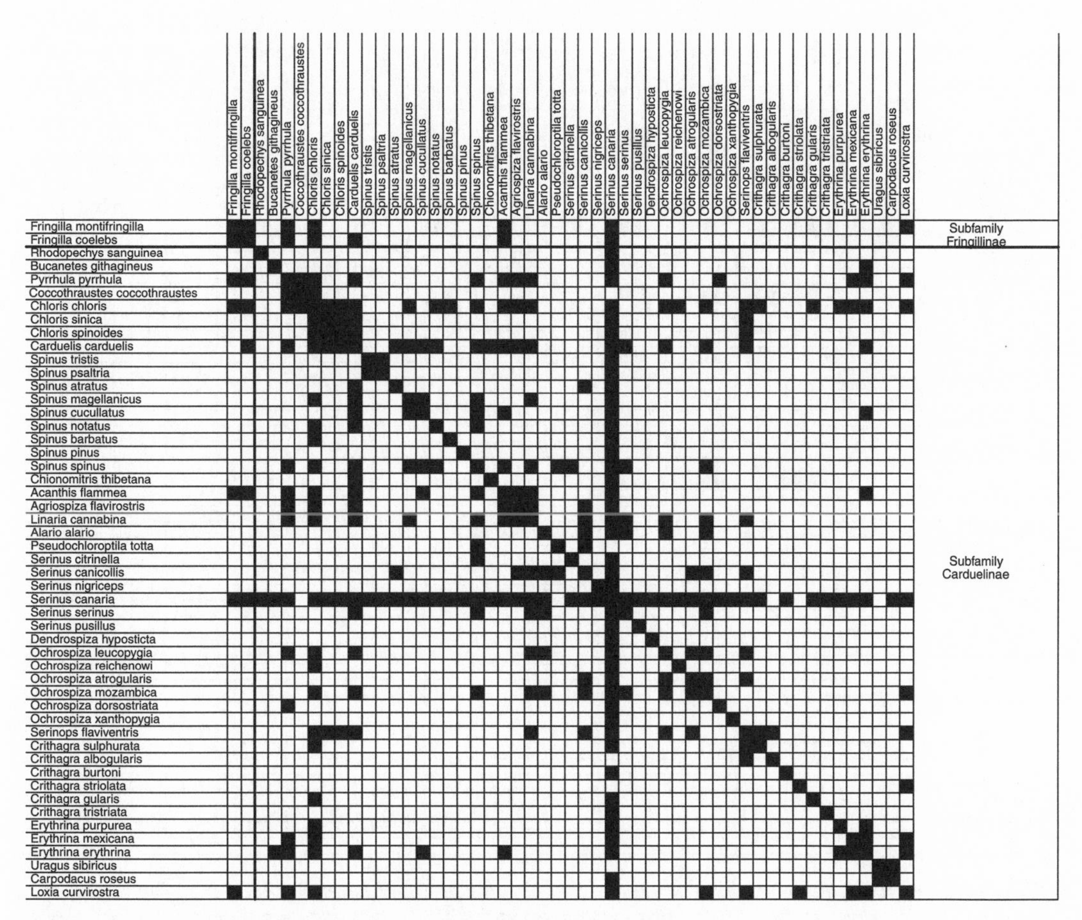

Figure 7.2. *A hybridogram for species of the finch family Fringillidae.*

hybrids do not exist for only five species, but in each case, the non-hybridizing species will cross with another species that can cross with the canary. Considering the canary's overall reproductive compatibility, we might conclude that any hybrids missing from the matrix are merely unreported rather than biological failures. Altogether, the canary alone establishes these representative members of the finch family as a monobaramin.

Because the grass family contains almost 10,000 species, the grass hybridogram displays successful hybrids between members of different grass tribes.[16] This hybridogram does two things to aid the viewer in recognizing continuity. The tribes are arranged in order from the largest tribe, Paniceae with 2,000 species, to the smallest tribes with only 1 species each. The ordering allows the viewer to recognize that the largest grass tribes also share the most continuity. Additionally, some of the cells in the hybridogram are grey rather than black, representing tribes that do not directly hybridize but that will hybridize with members of the same third tribe. In this way, we can infer from the grass hybridogram that approximately 7,200 grass species form a single monobaramin.

As seen in the previous illustrations, the hybridogram gives an immediate visual clue

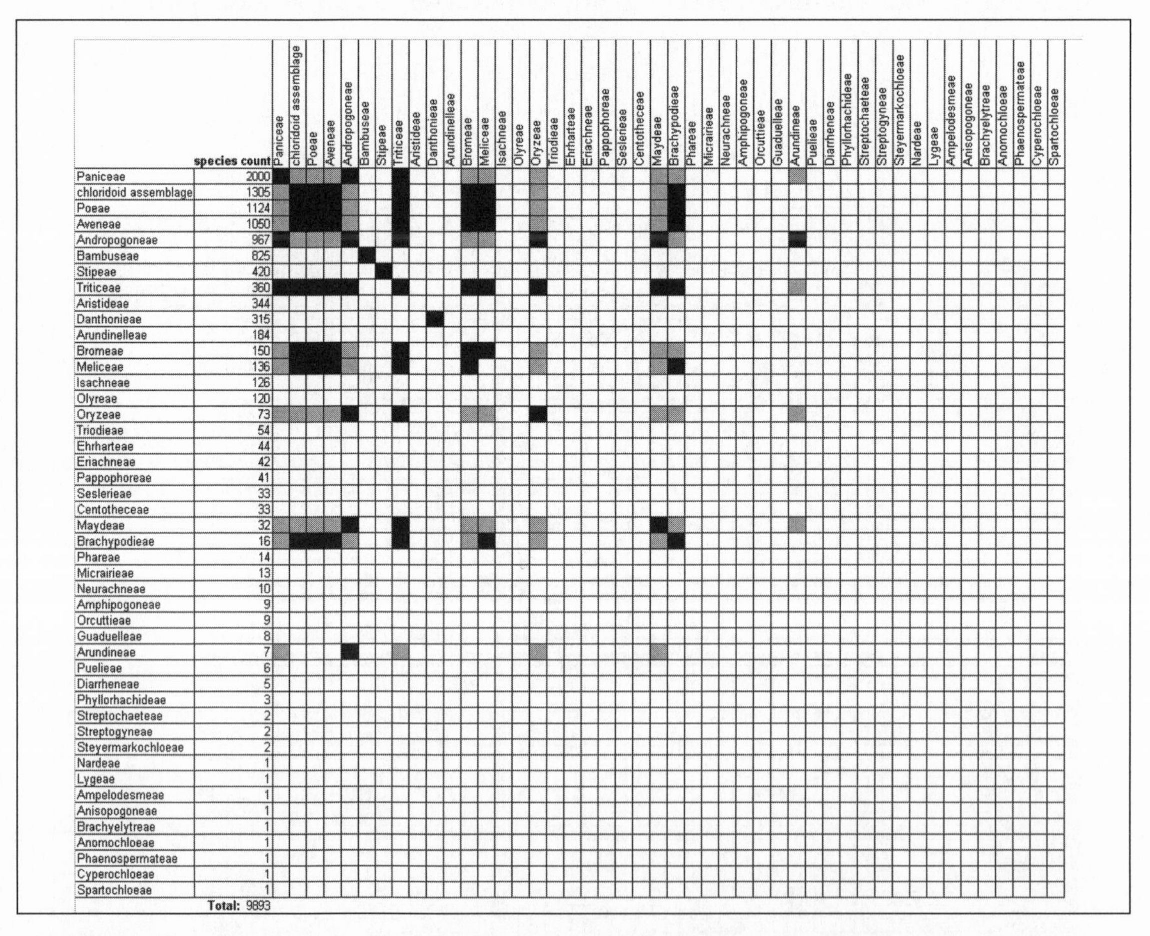

Figure 7.3. *A hybridogram for tribes of the grass family Poaceae. Tribes are ordered from the largest at the top to the smallest at the bottom. Tribes with members that can hybridize are shown as black squares. Tribes with members capable of crossing with the same third tribe are shown as gray squares.*

to the continuity of the group, but other evidence of continuity can supplement the information in the hybridogram. Robinson's turtle hybridogram shows not only turtle hybrids, but also other species of turtles that share higher DNA similarity than the hybridizing species.[17] From Robinson's hybridogram, we can infer two small turtle monobaramins, which both extend beyond the few species that actually hybridize. While DNA similarity by itself cannot establish continuity, combining it with hybridization can make it an important part of an argument for the existence of a monobaramin.

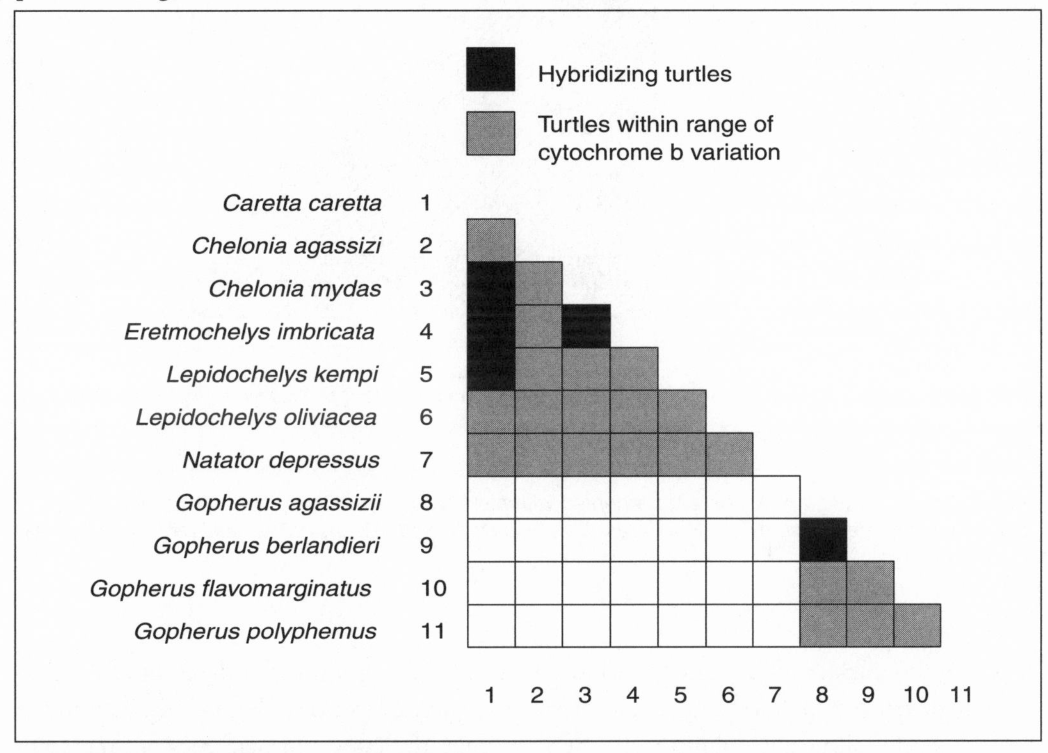

Figure 7.4. *A hybridogram of turtles supplemented with mitochondrial DNA similarity information. Turtles capable of hybridizing are shown as black squares. Turtles with DNA similarity greater than hybridizing turtles are shown in gray.*

Robinson and Cavanaugh use a similar approach in their hybridogram of extant cat species.[18] In this case, instead of using DNA sequence similarity, they use a statistical measure of the overall morphological similarity of the species (which we will present in the next chapter). Once again, they indicated successful hybrids by a black box and species within the range of similarity of the hybridizing species by grey boxes. From this hybridogram, we find that very few cat hybrids appear in the published literature, but by considering morphological similarity, we can extend the membership of the cat monobaramin to twelve of seventeen species.

7.6.2. Hybridization Network

While the hybridogram aids in the visual analysis of hybridization information, other graphical tools may better represent certain hybridization trends. For example, Scherer's

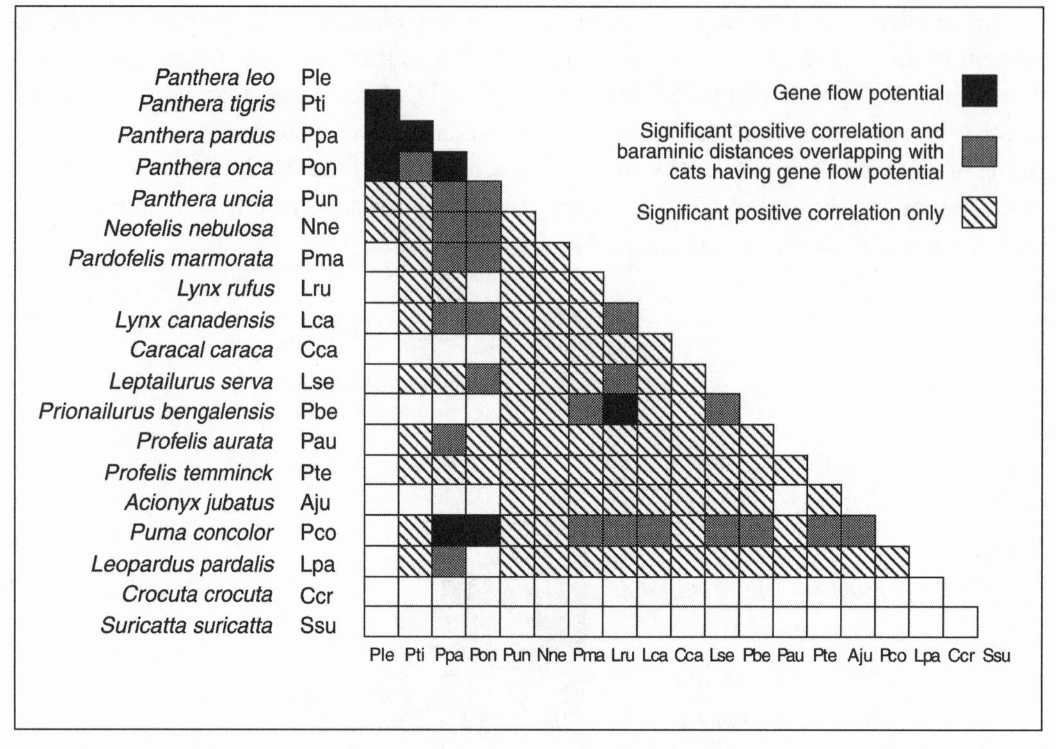

Figure 7.5. *A hybridogram of cats supplemented with morphological similarity data. As in Figure 7.4, hybridizing cats are shown as black squares. Cats with significant morphological similarity are shown as shaded squares.*

paper on the duck family Anatidae includes both a hybridogram and a diagram of the thirteen anatid tribes with lines connecting the tribes known to hybridize (Figure 7.6).[19] Although Scherer does indicate tribal limits in his hybridogram, the matrix format does not represent intertribal hybridization as well as the network diagram. From the tribal hybridization diagram, we can immediately see that the largest tribes in the family are all known to hybridize, whereas the smaller tribes with one to three species are not known to hybridize with other tribes. Based on this diagram, Scherer argues that the entire family probably represents a basic type (a monobaramin in our terminology).

For convenience, we will refer to such diagrams as **hybridization networks**. In a hybridization network, taxa are represented as nodes on a network where the connecting lines represent successful hybridization. Like the hybridogram, the hybridization network can be found in the conventional biological literature, used to represent trends in hybridization not easily represented in matrix format.

To construct an informative hybridization network, consider grouping the species of interest according to conventional classifications, prominent phenotypes, or other natural groupings. Begin by lumping species into genera or genera into tribes and construct hybridization networks in that way. Alternatively, if the group of interest could be classified

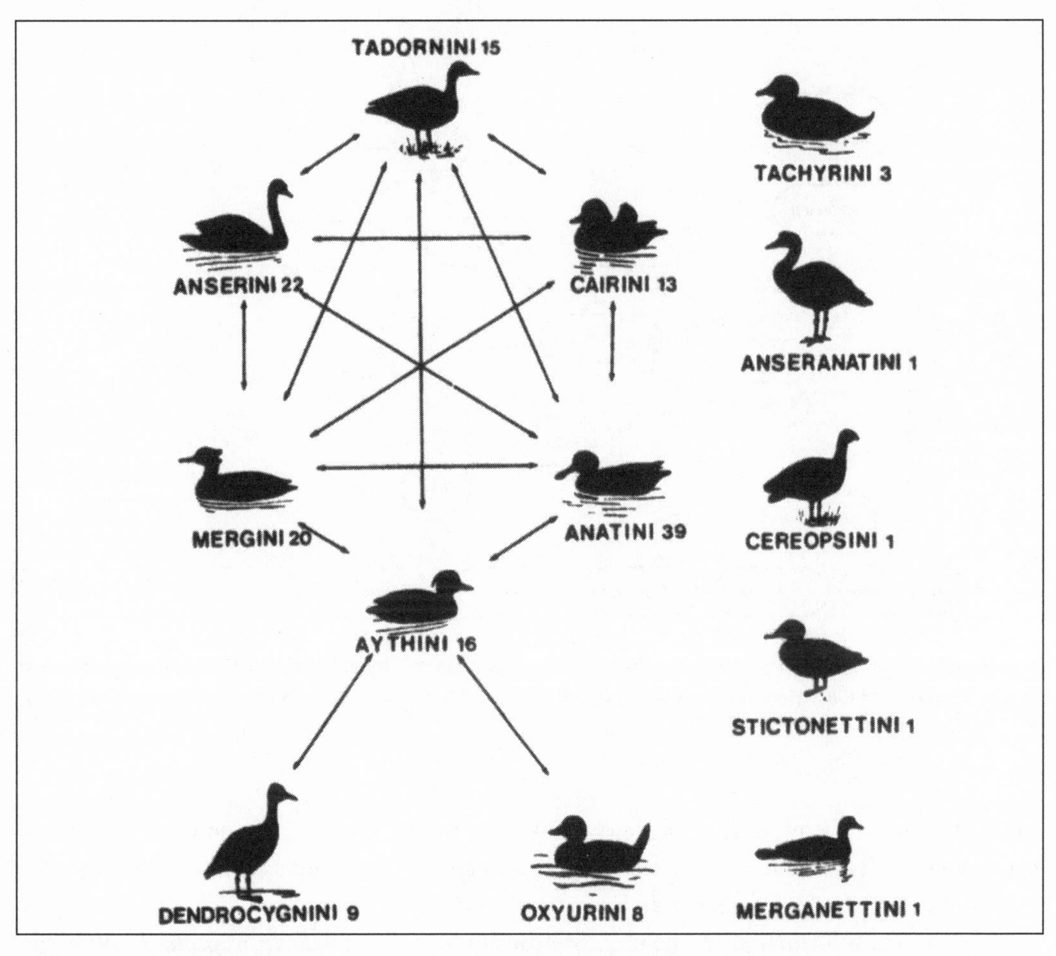

Figure 7.6. *A hybridization network for the duck family Anatidae. Birds are grouped according to their tribe, and hybridizing tribes are connected by lines.*

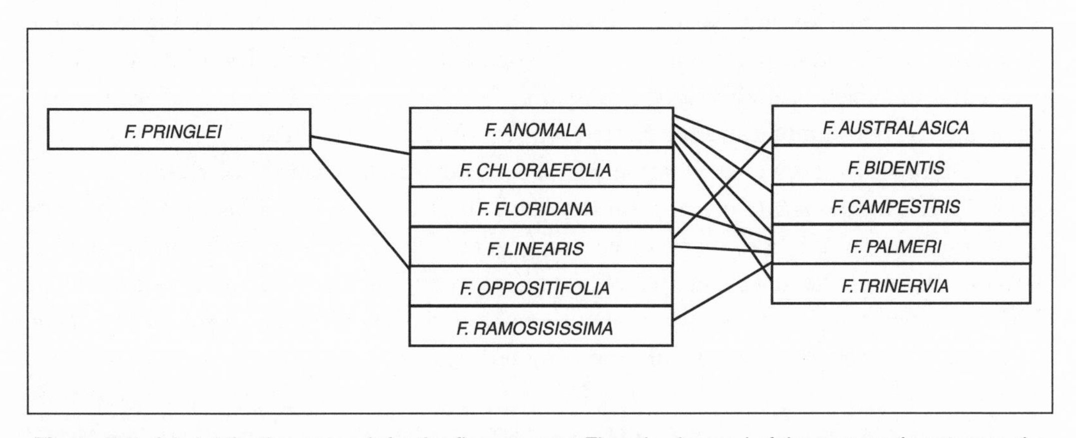

Figure 7.7. *A hybridization network for the flower genus* Flaveria. *Instead of the taxonomic category, the species are grouped by photosynthesis type. Hybridizing species are indicated by connecting lines.*

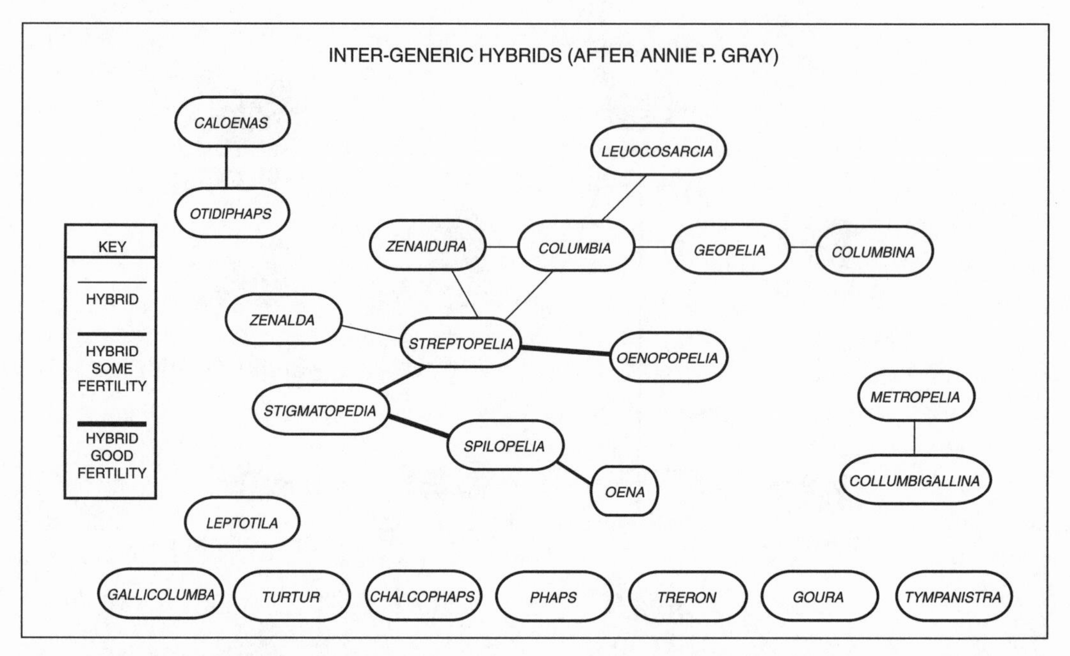

Figure 7.8. *A hybridization network for genera of the dove family Columbidae. Fertility of hybrid offspring is indicated by the thickness of lines connecting hybridizing genera. Thin lines represent sterile offspring, while thick lines represent fertile offspring.*

in another natural way, such as big and small cats or even groups from different geographical regions, a hybridization network may demonstrate important continuity between the groups that need to be emphasized.

To present information on the photosynthesis types of the yellowtop genus *Flaveria* (discussed in chap. 6), Wood and Cavanaugh created a hybridization network to illustrate the continuity between C_3, C_4, and C_3-C_4 members of the genus.[20] Rather than represent the types as nodes in the network, they merely grouped the species according to photosynthesis type and connected species capable of hybridization. The resulting network highlights the important hybridization trend that C_3 plants apparently do not hybridize directly with C_4 plants, but members of both will hybridize with C_3-C_4 intermediate species. Thus, the diagram effectively communicates the existence of continuity between the C_3 and C_4 species.

In some cases, additional information can also be displayed with a hybridization network. More's hybridization network for the doves illustrates fertility of intergeneric hybrids by the thickness of the lines that connect the genera.[21] In this way, we can readily see that eleven dove genera form a single monobaramin by hybridization, but more importantly, we find that a core group of genera within that monobaramin (*Stigmatopelia, Streptopelia, Spilopelia, Oenapopelia,* and *Oena*) are capable of producing fertile offspring.

7.7. Chapter Summary

Until recently, most non-creationist scientists have not recognized interspecific hybridization as the extensive biological phenomenon that it is. Some, however, have attempted to incorporate interspecific crossing into formal systematic definitions. Hybridization provides very powerful additive evidence when it is available and when it works. According to the refined baramin concept, the species of baramins and monobaramins share significant, holistic similarity (continuity). The successful production of hybrid offspring provides evidence of continuity because it involves the coordination of many levels of organization in the organism. Hybridization offers a practical method of demonstrating significant similarity between the genetics and development of two different species.

The high frequency of hybridization leads us to an important question: Why do we have species at all? We may infer from this curiosity that species must not be primarily reproductive units. Certainly reproductive isolation contributes to the present stabilization of species phenotypes, but reproductive isolation could not be the originator of species.

There are some practical resources that baraminologists can go to in order to discover hybridizations. Online resources include the HybriDatabase (HDB), the premiere source for hybridization information, specifically designed by and for baraminologists (http://www.bryancore.org/hdb/). Also, the National Center for Biotechnology Information offers free searching of thousands of abstracts related to biomedical research in their PubMed database (http://www.ncbi.nlm.nih.gov/pubmed/). Printed resources include *Mammalian Hybrids* and *Bird Hybrids* published by Annie Gray (both contained in the HDB). The Royal Horticultural Society (http://www.rhs.org.uk) keeps hybrid registers for *Clematis,* conifers, dahlias, larkspurs, *Dianthus,* lilies, daffodils, orchids, and rhododendrons. *Plant Breeding Abstracts* (PBA) and *Animal Breeding Abstracts* (ABA) are excellent printed resources.

Review Questions

1. Define interspecific hybridization.
2. How does hybridization demonstrate continuity?
3. Define hybrid swarm.
4. Define heterosis.
5. Define introgression.
6. List three practical problems with using hybridization in systematics.
7. List four ways to find hybridization information.
8. What problems led baraminologists to other baraminic membership criteria besides hybridization?
9. Why do scientists assume hybridization is rare?
10. Give an example of a sterile hybrid.
11. What two outcomes could result from introgression?
12. Define comparium.
13. Define syngameon.
14. Besides the HDB, what other resources can yield hybridization information?
15. Explain the advantages of a hybridogram.
16. Explain the advantages of the hybridization network.
17. How did More's hybridization network differ from Scherer's?
18. Define apomixis.

For Further Discussion

1. Why do we have species?
2. Explain the relationship between the comparium, syngameon, and Marsh's baramin?
3. Use the HybriDatabase to create a hybridogram for parrots (family Psittacidae).
4. Use the HybriDatabase to create a hybridization network for the genera of the bear family Ursidae.
5. Is it possible to overcome the practical pitfalls of hybridization? If so, how?
6. With all the limitations intrinsic in using hybridization information, why does it still enjoy a central position in baraminology?
7. Could the HybriDatabase ever be completed? If so, how? If not, why not?
8. What problems could arise when using a hybridogram? How could they be overcome?
9. What problems could arise when using a hybridization network? How could they be overcome?
10. How does one construct an effective hybridization network?

CHAPTER 8

Statistical Baraminology

8.1. Introduction

The first thing that novice baraminologists notice is that conventional systematics methods do not work for baraminology. All conventional systematics methods concern themselves primarily with identifying the historical path taken by organisms as they evolved. As a result, all conventional systematics methods *assume* that the organisms share an evolutionary ancestor. They offer no way to test the validity of common ancestry itself and consequently find limited application within baraminology. To identify continuity and discontinuity, several baraminologists developed novel techniques that we will discuss in this chapter. These techniques avoid making assumptions about the relationship of the taxa.

To review, a holobaramin is a group of species whose members share continuity but are discontinuous with all other species. Continuity and discontinuity are holistic terms, emphasizing that all characteristics of the organisms must be examined in order to discover significant similarity or differences. In part, we stress holism because so many modern systematics methods (such as cladistics) rely on very reductionistic, myopic evidence to define groups. For example, cladists might define a group based on a single characteristic, such as a particular skull protuberance. Unfortunately, creation biologists trained in this reductionist culture often have difficulty in breaking free and becoming holistic. As a result, creationists in the past pursued methods of baraminic identification that rely on a particular characteristic (such as hybridization potential) without examining other types of evidence.

We also emphasize holism because we believe it better represents God's plan of creation. During Creation Week, God made a mature world, with an appearance of age. Other passages of Scripture highlight God's value of maturity (1 Cor. 3:1–2). Since God created organisms as whole entities, we find it very unlikely that the baramins of which they are members ought to be defined or identified by such minuscule evidence as the similarity of a particular gene or the occurrence of a particular morphological trait. Because organisms were created whole, their baramins ought to be identified holistically.

In the previous chapter, we explained how hybridization provides evidence of continuity. The production of offspring requires coordination of many levels of organization (molecular, genetic, developmental, morphological); therefore, hybrids provide evidence of significant, holistic similarity (continuity). The use of hybridization in baraminology also has drawbacks, as we discussed in the previous chapter. When hybridization fails or cannot be used at all, we must find evidence of continuity in other ways, using careful data collection and sophisticated statistical methods. In this chapter, we will present two different methods that can be used to examine both continuity and discontinuity among species in a baraminology study. The baraminic distance method can provide statistical evidence of continuity and discontinuity.[1] **Analysis of Patterns** (ANOPA) provides the baraminologist with a method of viewing biological character space in three dimensions.[2]

8.2. Collecting Systematic Data

The baraminic distance and ANOPA methods rely on collection of useful baraminological data in the form of a data matrix. The data matrix consists of rows that represent species and columns that represent characteristics. In formal systematics, we call the species **taxa** (singular: **taxon**) and the characteristics **characters**. A character can be anything that describes a particular feature of an organism. For example, a height-to-weight ratio or the number of toes on the hind foot could both be characters. If a character could have an unlimited number of values, we refer to that character as continuous. Height/weight ratio is a continuous character, because no intrinsic feature of the character limits the number of values. A character that has a predefined number of values is called a discrete character. In fossil horses, the number of toes is a discrete character because it has a limited number of values possible.

The value of a character for a particular species is called that species' **character state**. For example, among cats, claws may be fully retractable or only partly retractable. In systematic terms, we would say that the character of claw retractability has two states: fully or partly retractable. The character state of the house cat is fully retractable, but the character state of the cheetah is partly retractable. Because cladistics deals only with discrete characters, systematists often focus their greatest attention on characters with limited states. The most convenient of the discrete characters are the binary characters, which have only two states, typically "present" or "absent." Discrete characters that have more than two states are called multistate characters.

For purposes of baraminic distance and ANOPA, the character state of a particular taxon is indicated in the data matrix by a numerical code. Binary characters are coded as "0" for absent and "1" for present. For multistate characters, the numerical coding goes as high as necessary to describe each of the states. While ANOPA can use both continuous and discrete characters, baraminic distance only allows discrete characters. Continuous characters may be converted for use in baraminic distance by classifying them into discrete groups such as quartiles. Sometimes, character states for particular taxa may be unknown or inapplicable, resulting in missing data. In baraminic distance, inapplicable characters or unknown states can be coded as "?" or "-." For ANOPA, all states must be numerical, so "0" is often reserved for missing data, with all other character states increased by one (e.g., 1 = inapplicable, 2 = absent, 3 = present, etc.).

Because baraminic distance and ANOPA examine the data holistically, baraminologists need not concern themselves with the same questions of data quality that plague the evolutionist. Cladistics treats character states as uniquely evolved attributes of organisms. For complex character states, it is probable that they evolved once and that species which possess them today inherited them from a common ancestor. If a character state arose more than once during the evolution of the group of interest, inclusion of these states in the analysis will muddle the phylogenetic inference.

Baraminologists may find it useful to examine the attributes of tree-like diversification models for their baramins, but baraminic membership can be established independently of such considerations. We make no assumption regarding the origin of complex character states; we only require that members of the same baramin share continuity with one another and discontinuity with everything else. Consequently, we may use a broad selection of characters, including behavioral, ecological, molecular, and morphological data, without any concern that we will violate the assumptions of our methods. We need only find continuity or discontinuity using a broad spectrum of evidence.

As much as possible, baraminologists should strive for a balanced and broad-based data matrix. For example, molecular sequence data have become popular among evolutionists for inferring phylogenetic history. As we mentioned in chapter 6, baraminologists ought to strive for a balanced, holistic dataset rather than one composed primarily of molecular characters. Since many hundreds of molecular characters (nucleotides or amino acids) can be derived from one or two gene sequences, inclusion of these characters without a similar number of other kinds of characters would violate the requirement of holism. Overloading a data matrix with molecular characters results in detection of false continuity between different baramins.[3] Similarly, morphological data that depend wholly on skull or tooth characters also violate the holistic requirement of the data matrix. Baraminologists should construct data matrices that include morphological characters from the entire body of the organism.

The most useful baraminological data matrices grow from the experience wrought by research with original material. Although researchers frequently publish data matrices from their own work (discussed below), these matrices often lack the holistic nature required by baraminology. Furthermore, using the data of other scholars subjects our results to their research interests. Because conventional biologists and systematists have no interest in baraminology, they will not necessarily produce matrices for organisms of interest to baraminologists or select characters that would be informative to baraminology. As a result, reinterpreting published data matrices places us at a disadvantage to the conventional systematist who generates his own data to any level of precision that he desires. Nothing substitutes for actual experience studying live or preserved specimens firsthand. In doing so, the baraminologist can create a data matrix uniquely tailored to the needs of baraminology and at the same time gain valuable experience that will aid in the interpretation of the results.

Although the importance of gathering firsthand data cannot be overemphasized, generating new data matrices also has disadvantages. Baraminologists may find it difficult to obtain access to specimens, particularly if they are not affiliated with an academic institution.

Since we must be good stewards with our time, we might better use our resources by examining data matrices in the published technical literature, even though they may not be ideal. In the past, baraminologists have focused almost exclusively on reinterpreting previously published data matrices, with good results. Considering the advantages and disadvantages of data collection from direct observation or publications, we recommend that baraminologists use the resources available to them wisely, first beginning with published information, then progressing as needed to primary observations and research.

Locating any published data matrix for your group of interest requires persistence. Not all systematists publish their data matrix with the results of their phylogenetic analysis, but a diligent search of the literature will usually reveal a data matrix for most prominent groups of plants or animals. Unfortunately, data matrices for fungi or unicellular organisms are much harder to obtain, since nearly all systematists now use DNA similarity to study such groups. In such cases, baraminologists have no choice but to create their own data matrix from either published observations or original research.

Two internet databases provide a repository for the data matrices from published systematics papers. In both cases, the databases are fairly young, and authors only contribute data matrices voluntarily. As a result, both databases contain few data matrices. The University of Buffalo provides data matrices in their TreeBase (http://www.treebase.org/treebase/index.html). Alternatively, the Cladestore database (http://palaeo.gly.bris.ac.uk/cladestore/default.html) from the University of Bristol can also be a source of data matrices. Both of these databases can be searched for formal names of taxonomic groups, and the data can be downloaded in a number of formats.

When these sources do not provide the data needed for a baraminology study, consult specialty publications, such as the journals *Cladistics* or *Systematic Biology*. Journals like these focus exclusively on systematics and frequently publish data matrices to accompany their articles. Alternatively, journals that publish papers on specific groups of organisms (like *Journal of Mammalogy* or *Herpetologica*) sometimes publish systematics articles with accompanying data matrices. Also remember to consult the literature databases mentioned in the previous chapter. Search for keywords *systematics* or *cladistic* to identify appropriate papers that might contain data matrices. If a recent systematics paper on the group of interest contains no published matrix, consider contacting the senior author directly and politely requesting the data matrix.

Prior to ANOPA calculations, all character states must be normalized for each taxon. Normalized character state values are calculated by the following equation:

$$CS'_{ki} = \frac{CS_{ki} - \min(CS_i)}{\max(CS_i) - \min(CS_i)}$$

[8.1]

where CS'_{ki} is the normalized character state for character i and taxon k, CS_{ki} is the actual character state value for character i and taxon k, and $\max(CS_i)$ and $\min(CS_i)$ are the maximum and minimum character state values for character i. Normalization scales the values from 0 to 1, allowing ANOPA to analyze any mixture of discrete binary characters, discrete multistate characters, and continuous characters. Baraminic distance calculations do not require normalized data.

8.3. Baraminic Distance

8.3.1. The Method

To a conventional biologist, the word *distance* when used in systematics refers to any of several kinds of numerical summaries of the differences between organisms. Often, these distances themselves are calculated based on the assumption of common ancestry, especially in sequence analysis. If we intend to test the hypothesis of common ancestry, baraminologists need a distance measurement that is unfettered by assumptions of ancestry.[4]

Baraminic distance provides just such a measure. Developed by Robinson and Cavanaugh, the baraminic distance technique also provides a number of statistical measures that summarize the overall quality and completeness of the data. The baraminic distance is simply a percent difference between the character states of two taxa, calculated as follows:

$$d_{ij} = \frac{m_{ij}}{n_{ij}}$$

[8.2]

where d_{ij} is the baraminic distance between taxon *i* and taxon *j*, m_{ij} is the number of characters for which the state of *i* is different than the state of *j*, and n_{ij} is the total number of characters compared for taxon *i* and *j*. Note that we do not include missing characters when calculating m_{ij} and n_{ij}.

According to our refined baramin concept, continuity and discontinuity are significant, holistic similarity and difference within biological character space. The baraminic distance estimates the multidimensional distance D_{ij} between two taxa in character space. Since we would have to measure every possible character on taxa *i* and *j* to know their true distance, we try to estimate it instead with a holistic d_{ij} measurement. Consequently, the more characters we add to our data matrix, the closer we will come to an accurate estimate of D_{ij}.

Because so many data matrices include incomplete character states, Robinson and Cavanaugh also introduced diagnostic measures to evaluate the completeness of the data. The **character relevance** is the percentage of taxa for which a state is known, calculated as follows:

$$a_i = \frac{x}{n_t}$$

[8.3]

where a_i is the relevance of character *i*, *x* is the number of taxa for which a state for character *i* is known, and n_t is the total number of taxa. Baraminologists should prefer characters with high relevance to make sure that the information in the data matrix will adequately represent the true distance between the taxa. Robinson and Cavanaugh recommend that characters with less than 95% relevance be eliminated from baraminic distance calculations.

To understand the information content of the data matrix, Robinson and Cavanaugh introduced **character diversity** as a measurement of the variation in the matrix. Diversity is calculated according to formula [8.4]:

$$c_i = 1 - \sum f_{ij}^2 \left(\frac{n_t}{n_t - 1} \right)$$

[8.4]

where c_i is the character diversity for character *i*, f_{ij} is the frequency of character state *j* for character *i*, and n_t is the total number of taxa. Diversity approximates the probability that two taxa selected at random will differ at character *i*. Low character diversity should be avoided, but higher diversity does not necessarily mean better data. If all characters differ for all taxa, diversity would be high, but information content would be low (every taxon would be unique). If taxa differ only by one or two characters, then diversity and information content would be low (all taxa would be nearly the same). Good diversity is moderate diversity.

Even though the character relevance may be high, and the character diversity may be moderate, we still might not have a baraminologically informative data matrix. To evaluate the uniformity of the data, Robinson and Cavanaugh also introduced **baraminic signal** (*SI*). Based on the chi-square statistic, *SI* is calculated after all distances have been calculated as follows:

$$SI = \sum \frac{(d_{ijOBS} - d_{ijEXP})^2}{d_{ijEXP}}$$

[8.5]

where d_{ijOBS} is the observed baraminic distance (equation [8.2]) and d_{ijEXP} is the expected baraminic distance, calculated as a standard expected value in a chi-square:

$$d_{ijEXP} = \frac{(\sum_{i=1}^{n_t} d_{ijOBS}) \times (\sum_{j=1}^{n_t} d_{ijOBS})}{\sum_{i=1}^{n_t} \sum_{j=1}^{n_t} d_{ijOBS}}$$

[8.6]

Or in simpler terms, d_{ijEXP} is the product of the sum of the observed distances for taxon *i* by the sum of the observed distances for taxon *j* divided by the total distances of all organism pairs. A low baraminic signal indicates a uniform set of taxa that are roughly equidistant from each other. A high baraminic signal implies the presence of two or more distinct clumps of organisms.

Once the distances for each pair of organisms have been calculated, they can be used in any number of ways to demonstrate significant similarity or difference. The distribution of the ingroup distances could be compared to the distribution of ingroup/outgroup distances to determine the statistical significance of any differences. We recommend that baraminologists use nonparametric statistics methods for comparing baraminic distance distributions. Tests such as the standard t test make assumptions about the data distributions which are probably incorrect for baraminic distances. A Wilcoxon rank sum test would be a better choice.

As an alternative method of detecting significant similarity or dissimilarity, Robinson and Cavanaugh introduced a distance correlation test, which they subsequently used most frequently. By calculating a standard linear correlation coefficient for all the distances involving either of two taxa *i* or *j*, baraminologists can detect either significant similarity or significant difference between taxa *i* and *j*. The method works because two taxa close together in character space will have similar distances to outlying taxa, whereas two taxa that lie very far apart in character space will have an inverse distance relationship. For taxa far apart, taxa close to one will be far from the other and vice versa. Thus, by calculating a linear correlation coefficient between the distances of two taxa to all other taxa, we can determine if the taxa are close together or far apart. Since statistical significance can be calculated for linear regression models, we can also determine if two taxa are significantly close or significantly distant by using the same method.

At this point, we must call attention to the fact that significant difference or significant similarity discovered in baraminic distance distributions only *suggest* discontinuity or continuity, respectively. Because continuity or discontinuity properly describe holistic character space, any limited, finite dataset will necessarily only sample some dimensions of character space. As a result, d_{ij} only approximates the true distance D_{ij} as we noted above. Therefore, significant similarity or difference discovered in this finite data matrix can only suggest continuity or discontinuity. It is possible that significant similarity between baramins could be demonstrated with a very general data matrix or that significant difference could be detected within a very diverse baramin. In no way does this invalidate baraminology, since baraminology works by approximation.

To stimulate baraminology research using the baraminic distance method, Wood developed software that reads a data matrix and calculates baraminic distance, character relevance, and character diversity.[5] Called BDIST, the program first calculates character relevance (a_i) and automatically eliminates characters with a_i<0.95. The program then calculates the character diversity for the remaining characters. Both relevance and diversity are reported in the output of BDIST. Using the remaining characters, baraminic distances are calculated and presented in a matrix format that can be imported readily into other software, such as a spreadsheet or statistics package. BDIST is distributed free of charge at the internet site of the Baraminology Study Group (http://www.bryancore.org/bsg/). The site also includes a detailed tutorial for using BDIST.

8.3.2. Baraminic Distance Applications

Because baraminic distance was only published in 1998, few studies using the method have appeared in the technical literature at the time of this writing. The studies that have been done have focused on a wide variety of groups, including animals, plants, and fossils. Robinson and Cavanaugh originally applied the method to primates, illustrating that morphological data better resolved discontinuity among the primates than sequence data.[6] They followed up their first study with an analysis of the cats (discussed in chap. 5),[7] in which they demonstrated significant similarity among extant cats and significant differences between cats and non-cat carnivores. Consequently, they proposed that the cat family Felidae was a holobaramin. Wood used the method to study the grass family Poaceae,[8] and Wood, Cavanaugh, and Wise used the method in their analysis of fossil horses.[9]

8.3.2.1. Catarrhine Primates. The first use of baraminic distances appeared in Robinson and Cavanaugh's analysis of the baraminology of the primate suborder Catarrhini, including families Cercopithecidae (Old World monkeys), Hylobatidae (gibbons), Pongidae (gorillas and chimps), and Hominidae (humans). This group of organisms offers an excellent internal control in the human/ape relationship. Since we know that humans are not continuous with other organisms (Gen. 1:26–27; 2:7), we must be able to find significant, holistic differences between humans and apes. Furthermore, Hartwig-Scherer proposed that the family Cercopithecidae forms a basic type (monobaramin), based on hybridization potential.

Robinson and Cavanaugh developed their own data matrix from the published literature, consisting of 204 characters from 11 different species. The characters covered ecological (18 characters), morphological (43 characters), and molecular information (143 characters, which they divided into two categories, sequence and chromosomal). Because this matrix is skewed toward molecular information (70 percent of the characters), we should not label the matrix holistic, but Robinson and Cavanaugh divided the matrix into character categories and analyzed them separately to study the usefulness of different categories of information.

Two examples of their baraminic distance scatterplots are shown in figure 8.1. In the first, the baraminic distances of the mandrill are plotted with the baraminic distances of the baboon. From Hartwig-Scherer's analysis, we know that these two species belong to the same monobaramin because they can successfully hybridize.[10] The scatterplot of their baraminic distances confirms this relationship by demonstrating significant positive correlation with a correlation coefficient of 0.891. In contrast, the baraminic distances of the mandrill and the baraminic distances of the human correlate negatively, with a correlation coefficient of -0.761. This result is consistent with the biblical data that reveals that humans and animals have separate origins and do not belong to the same baramins.

Using morphological or ecological characters, the human holobaramin and the non-human primates exhibited significant differences. This result confirmed that baraminic distances could detect significant differences when true discontinuity was present. In contrast, the molecular sequence data did not reveal significant differences between the humans and non-human primates, indicating that sequence data may not be useful for detecting discontinuity. Our refined baramin concept requires that baraminological analyses be done holisti-

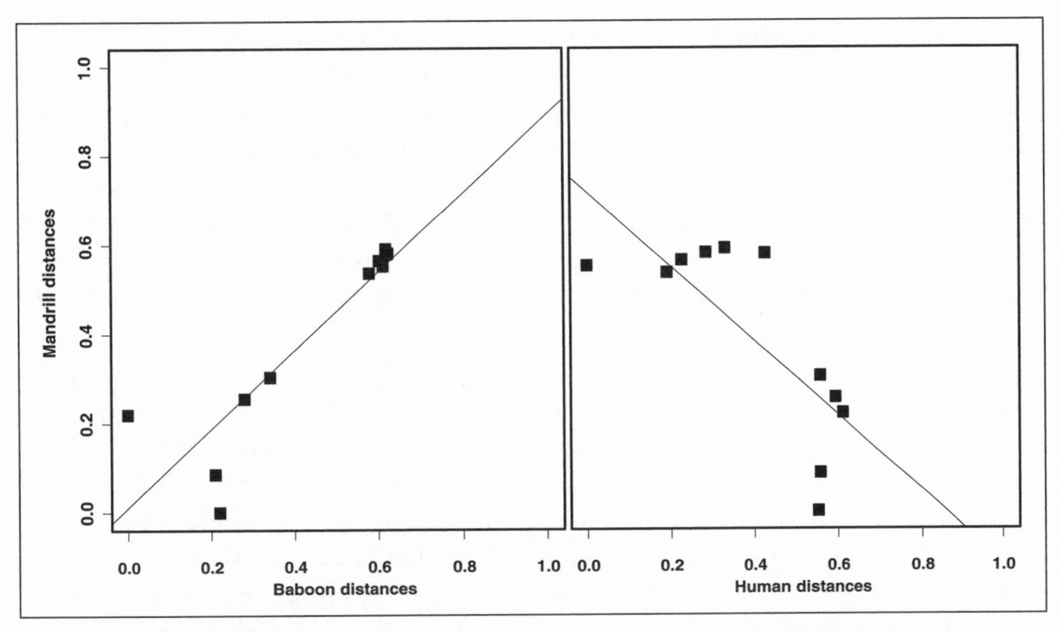

Figure 8.1. Baraminic distances of the mandrill and baboon correlate positively (left), indicating significant similarity between the two taxa. In contrast, the baraminic distances of the mandrill and human correlate negatively (right), indicating significant difference between the two taxa.

cally to represent the character space more accurately. Since 114 of the molecular sequence characters came from a single gene (12S rRNA), we do not find it surprising that these characters failed to distinguish truly discontinuous groups. The more holistic data in the morphological and ecological characters accurately represented character space and the true discontinuity.

8.3.2.2. Grasses. To introduce his BDIST software, Wood performed a baraminic distance analysis of the grass family, Poaceae. Poaceae contains nearly ten thousand described species, including such economically important crops as maize, wheat, rice, barley, rye, sugarcane, and bamboo. Because of its importance to humans, Poaceae has become a well-studied group, and even some creationists have studied these interesting plants. Junker analyzed one tribe of the family (Triticeae) and declared it a basic type (monobaramin).[11] Ongoing research at the Van Andel Creation Research Center in Prescott, Arizona, focuses on ring muhly, an unusual desert grass species.[12]

A number of grass specialists recently formed the Grass Phylogeny Working Group (GPWG) to study the evolutionary history of the family. The GPWG has compiled a data matrix with 7,025 characters for 66 genera (62 grass, 4 outgroup), which is available for download from their website (www.virtualherbarium.org/grass/gpwg/). The 62 grass genera represent 36 of the 46 recognized tribes. Because the characters consist of 53 morphological characters and 6,972 molecular and chromosomal characters, we should expect that the full data matrix will not reveal significant differences between any of the taxa. The 53 morphological characters alone should provide a more holistic view of the baraminic status of these species.

Wood's results confirmed that the full data matrix of 7,025 characters revealed significant similarity between grasses and the four non-grass outgroup taxa. Only 11 of the 62 grass taxa displayed significant differences with any one of the outgroup genera. When he examined the 53 morphological characters separately, he found that nearly every grass genus was significantly different from each outgroup genus. To his surprise, he also found two grass genera, *Streptochaeta* and *Anomochloa,* that were significantly similar to the outgroup taxa and significantly different from other grass species. A third grass genus, *Pharus,* displayed little similarity to other grasses and was significantly similar to one of the outgroup taxa.

We see this correlation in the scatterplots of figure 8.2. For all of these plots, only the baraminic distances using the morphological data are shown. In the first panel, the baraminic distances of wheat and rice are shown. These two grass species are significantly correlated with a correlation coefficient of 0.662. In contrast, when the wheat distances are plotted with the distances of the non-grass *Elegia,* we find a significant negative correlation, with a correlation coefficient of -0.761 (next panel). Despite being classified in the Poaceae, *Streptochaeta* correlates negatively with other members of Poaceae, such as wheat. The correlation coefficient of the wheat/*Streptochaeta* plot is -0.659. In the last panel, we see the baraminic distances of *Streptochaeta* and *Elegia,* which share significant positive correlation, with a coefficient of 0.342.

Based on the results of the morphological baraminic distance correlations, Wood proposed that the grass family Poaceae was a holobaramin because its members shared significant similarity with one another and significant difference with outgroup genera. He excluded *Streptochaeta* and *Anomochloa* from the Poaceae holobaramin because of their significant differences with many other grasses and their significant similarity with the outgroup. To determine the baraminic status of *Pharus,* he suggested that further study would be necessary.

8.4. Analysis of Patterns (ANOPA)

According to the refined baramin concept, baramins and potentiality regions exist in a multidimensional biological character space. As three-dimensional creatures, we cannot perceive multidimensional space. Because of this perceptual limitation, most baraminological methods attempt to approximate the features of multidimensional space by measuring baraminic distances or by demonstrating genetic continuity via hybridization. While these methods have proved very helpful in the past, a means of observing and studying taxa in multidimensional space would be ideal for examining baramins and potentiality regions. Although no such method exists to allow us to observe multidimensional space directly, the **Analysis of Patterns** method devised by Cavanaugh comes the closest.

Analysis of Patterns (ANOPA) takes multidimensional data and reduces the dimensionality while minimizing the loss of information. ANOPA results can be examined in one, two, or three dimensions, with 3D ANOPA containing the most information about the taxa. By examining the patterns in 3D ANOPA, baraminologists can formulate baraminic hypotheses based on a direct approximation of multidimensional character space.

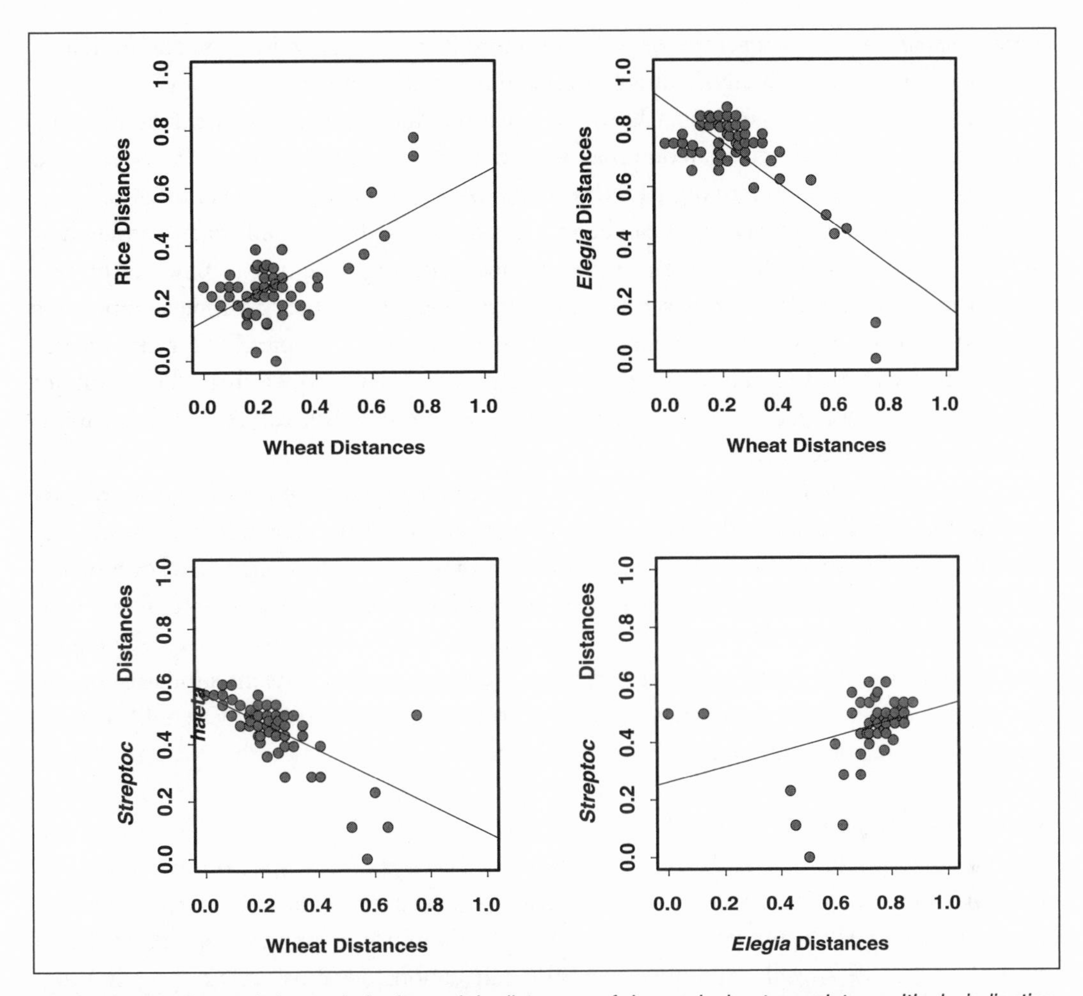

Figure 8.2. *In the top left panel, the baraminic distances of rice and wheat correlate positively, indicating significant similarity between the taxa. In the top right panel, the baraminic distances of wheat correlate negatively with the baraminic distances of the non-grass genus* Elegia. *Surprisingly, the baraminic distances of the grass genus* Streptochaeta *correlate negatively with the baraminic distances of wheat, a fellow grass. The baraminic distances of* Streptochaeta *and* Elegia *have a slight positive correlation, indicating possible continuity between these taxa.*

ANOPA also has a tremendous advantage over conventional systematics methods, such as cladistics or phenetics. Conventional phylogenies force the data to adopt a particular geometric structure, namely a bifurcating tree. Naturally, this limits the information that can be depicted. In particular, biological similarities that do not agree with the majority of characters cannot be represented on the evolutionary tree. Baraminologists recognize that the pattern of life need not be tree-like, and ANOPA provides a means of observing non-tree patterns in character space.

ANOPA resembles other multidimensional analyses methods commonly used in systematics research, but numerous distinctions set ANOPA apart. Discriminant analysis can be useful for classification of objects if some *a priori* knowledge about the true classification is available. Since ANOPA makes no assumptions about the geometric structure of the taxa in character space, it can be useful where the structure is unknown, as in most baraminological studies. Discriminant analysis thus has limited applications in baraminology.

Principle component analysis (PCA) attempts to reduce multidimensional data by capturing variance in linear vectors (components). Unlike discriminant analysis, PCA makes no assumptions about the underlying structure of the data examined, but the amount of information captured in the components can be very limited for highly variable data matrices. For uniform populations, the first three principle components might capture as much as 85 percent of the variance in the full dataset. For highly variable groups, the first three components can capture very little variance, sometimes less than 50 percent. Typically, if a data matrix varies in many dimensions at once with little to no correlation between them, PCA will not recover that variance well. As a purely descriptive method, ANOPA can recover more information than the variance-dependent PCA.

The strongly descriptive nature of ANOPA can also be a weakness if not used carefully. With hybridization and baraminic distances, we could demonstrate statistically significant, or at the very least holistic, similarity or difference between groups of organisms. By itself, ANOPA only provides a depiction of what the taxa might look like if we could perceive character space. We then must apply other statistical methods or baraminological techniques to the ANOPA results in order to detect traces of significant similarity or difference. As long as baraminologists remember to use ANOPA carefully, it could become one of our most powerful tools.

8.4.1. The Method

8.4.1.1. The Centroid. ANOPA treats taxa (individual organisms, species, genera, etc.) as points in multidimensional space, wherein each character represents a separate dimension. The character states for any given taxon define a position for that taxon in multidimensional space. As a first step for all subsequent ANOPA calculations, we must calculate a point that will serve as our reference for the taxic group. Termed the *t-group centroid,* this hypothetical, "average" taxon eliminates the bias of using a particular taxon as a reference, allowing taxic patterns to be more detectable. The centroid point is calculated as the average character state for each character over all taxa.

8.4.1.2. One-dimensional (1D) ANOPA. The first method of drawing patterns out of the dataset is the calculation of the Euclidean distance from each taxon to the centroid, as follows:

$$a0_k = \sqrt{\sum_{i=1}^{n_C} \left(CS'_{ik} - CS'_{icentroid}\right)^2}$$

[8.7]

where $a0_k$ is the 1D ANOPA distance for taxon k, n_c is the total number of characters, CS'_{ik} is the character state value for character i and taxon k, and $CS'_{icentroid}$ is the character state value of character i for the centroid (the average character state value for character i). We can think of the $a0$ distance as the deviation of each taxon from the norm of all taxa in the data matrix.

By plotting the *a0* distances as a histogram, we can get an initial picture of the grouping and separation of the taxa. As an alternative, a cumulative distribution displays the sorted *a0* values plotted against the empirical rank or normalized rank of each *a0* value. While the cumulative distribution plot provides information about the precise placement of each individual taxon, the histogram can aid in identifying clumps of data. Figure 8.3 demonstrates the difference between the histogram (top) and the cumulative distribution plot (bottom) for the same set of 100 hypothetical *a0* distances. In the figure, we can see at least three major populations in the cumulative distribution plot, as well as an outlying taxon. In the histogram, we see more clearly that the first population, from $a0 = 0.15$ to 0.5, actually appears to consist of two overlapping subpopulations.

8.4.1.3. Two-dimensional (2D) ANOPA. To continue the transformation of the multidimensional character data into a three-dimensional projection, we need to select an outlying taxon or group of taxa from the *a0* plot. A hyperline (multidimensional line) is then drawn from the centroid to the outlying taxon. In the case of a clearly defined outlying group of taxa, a centroid of just those taxa may be calculated and used as the outlying point. For convenience, we will refer to the outlying point or taxon as simply the "outlier." Using the hyperline between the centroid and outlier, two numbers are calculated for each taxon. First, the *t0* distance is calculated as the distance of each taxon from the centroid along the vector of the hyperline. Second, the perpendicular distance from the hyperline to each taxon point is calculated as the *d2* distance. These two distances form the *x* and *y* coordinates for the two-dimensional ANOPA plot.

The 2D ANOPA results can be plotted in two different ways to reveal underlying clustering patterns in the biological data. First, the *t0* statistics can be plotted as a cumulative distribution, similar to the 1D *a0* statistics. Whereas the *a0* statistics measure the Euclidean distance from each taxon to the centroid, the *t0* statistic measures the distance from the centroid to the taxon along the pattern vector. By examining just the *t0* distribution, clustering patterns in addition to the patterns of the *a0* distribution can be detected. Ideally, though, the *d2* and *t0* statistics should be plotted as a 2D scatterplot, which will reveal clusters and patterns not detectable in the one-dimensional *a0* or *t0* plots.

8.4.1.4. Three-dimensional (3D) ANOPA. Taken by itself, a scatterplot of the *d2* and *t0* distances can provide a better view of the diversity of the taxa than a simple 1D ANOPA *a0* plot, but three-dimensional (3D) ANOPA provides the best projection and separation of the multidimensional character data. The three dimensions of the 3D ANOPA are cartesian coordinates derived from the multidimensional biological character space. The cartesian coordinates for each taxon are converted from the polar cylindrical coordinates that originate with reference to the hyperline between the centroid and the outlier. The *d2* and *t0* statistics from the 2D ANOPA provide the distance measurements for the cylindrical coordinates, and the angle formed by the taxon, the hyperline, and the origin of the multidimensional space provides the third statistic for calculating the cartesian coordinates.

For viewing 3D ANOPA, Cavanaugh recommends using the 3D viewing software Mage (http://kinemage.biochem.duke.edu/), originally designed for viewing biochemical structures. Mage lets the user rotate and translate the 3D ANOPA points to view the results as well as position the points for generating images for publication. Individual data points in

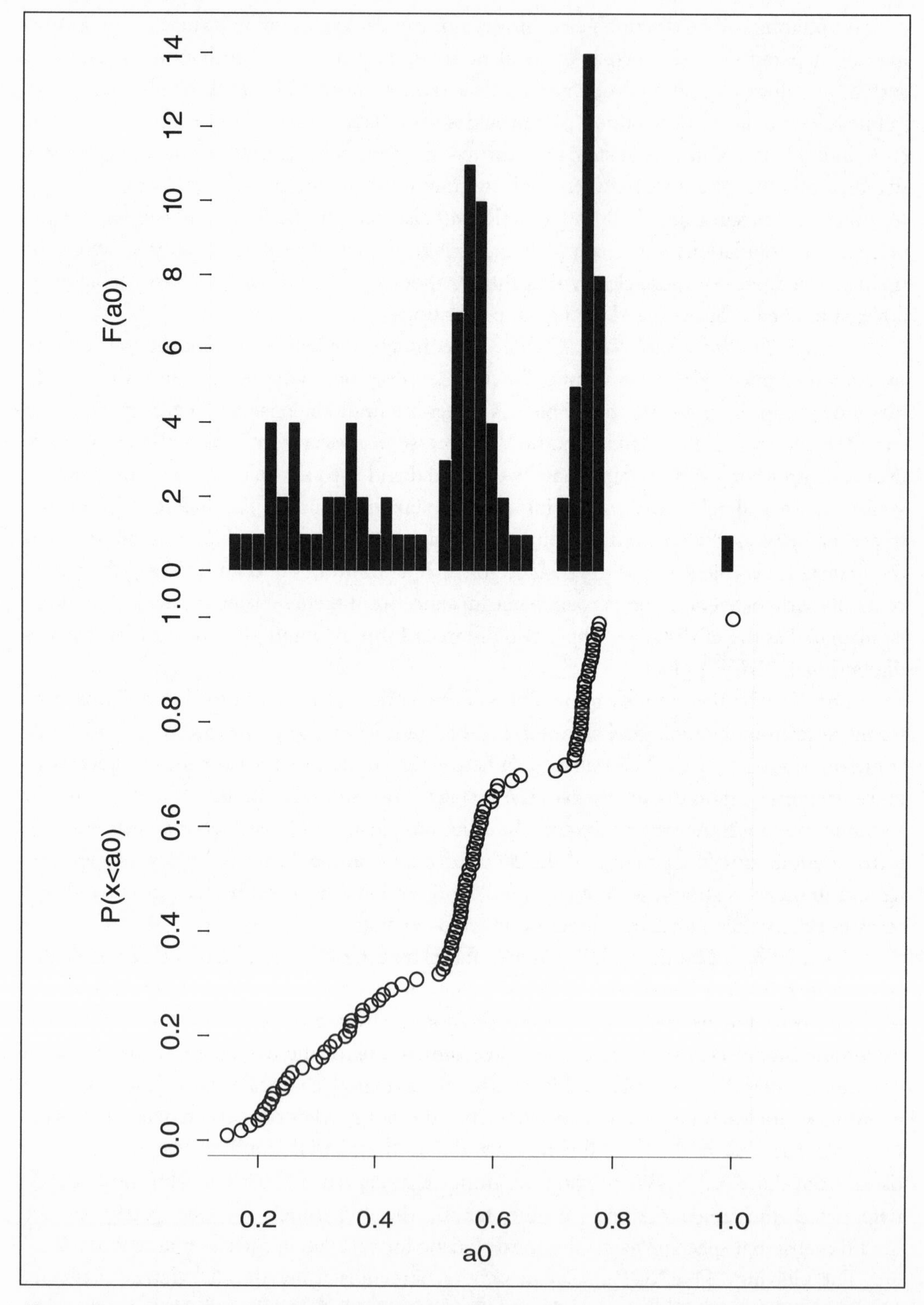

***Figure 8.3.** A population of randomly generated* a0 *distances plotted first as a histogram (top) and as a cumulative distribution (bottom). Each perspective has advantages for visualizing ANOPA distributions.*

Mage can be identified by altering their color or by labeling them. Mage can output a single two-dimensional projection of the three dimensional data or a "stereo" image. Mage outputs graphics in PostScript format, which can be imported into CorelDraw or other Corel software products (such as WordPerfect or Presentations) for additional editing.

8.4.2. ANOPA and Conventional Classification

Because conventional classification limits itself largely to bifurcating tree structures, informative taxonomic data is necessarily lost when fit to a tree. At the same time, systematists very rarely use non-tree methods of classification such as PCA or reticulistics. As a descriptive method that considers all the data, we should expect that taxonomic clustering patterns identified by ANOPA might disagree with phylogenetic classifications employed by conventional systematists. ANOPA patterns can–and do–exhibit a wide variety of configurations, from compact globular structures to complicated arcs. For some groups, these patterns might agree with a conventional classification, while other groups show a clustering in ANOPA markedly different than the phylogenetic tree for the same taxa.

In this section, we will present two groups of organisms that illustrate the conflict between ANOPA and conventional classifications that baraminologists should expect to encounter. The sunfish family Centrarchidae typifies a small group of organisms with a poorly defined phylogeny, and ANOPA has aided in clarifying the relationship of these fish. Because no outgroup taxa were included in the sunfish study, no baraminological inferences can be made based on the results. In the second example, Wood and Cavanaugh examined the Heliantheae tribe of the sunflower family Asteraceae using ANOPA. They included members of an outgroup tribe (Eupatorieae) in their analysis to identify potential evidence of discontinuity. They found that the ANOPA clustering patterns significantly differed from the conventional classification.

8.4.2.1. Sunfish.[13] Cavanaugh and Sternberg focused on the fish family Centrarchidae, comprising the basses, sunfish, and crappies, for their ANOPA study. This North American family consists of twenty-seven species, with only one (Sacramento perch, genus *Archoplites*) occurring west of the Rocky Mountains. Despite numerous attempts to study centrarchid phylogeny, no one has discovered an evolutionary tree that completely accounts for the morphological data of the sunfish. Beginning with a published data matrix of fifty-three characters for twenty-two species of the family, Cavanaugh and Sternberg evaluated the taxonomic groupings of the Centrarchidae by using ANOPA.

Even in the 1D ANOPA of the centrarchids, five clusters of species are detectable from the *a0* histogram. The cumulative distribution of the *t0* distances support these five groups. At this level, most Centrarchid species cluster together with other species of their genera, with the exception of the species *Lepomis gulosis,* which occurs alone in a transitional position between the genera *Ambloplites* and *Acantharchus*. In contrast, the other members of the *Lepomis* genus appear very closely clustered at a *t0* distance of 0.2 from *L. gulosis.*

The two-dimensional (*t0, d2*) plot confirms the isolated position of *L. gulosis,* but with the added resolution afforded by the *d2* dimension, we can see that *L. gulosis* is a transition between *Ambloplites* and *Enneacathanus,* rather than *Acantharchus*. In the 3D ANOPA, the centrarchid family forms an arc shape with *L. gulosis* located in the interior of the arc. The same five clusters of species observed in the one- and two-dimensional ANOPA plots also appear in the 3D distribution.

Based on analyses of the variation of the 2D ANOPA results, Cavanaugh and Sternberg concluded that the Centrarchidae form a statistically significant group, and we would add that these fish might belong to the same monobaramin based on these results. Because they did not include an outgroup in their analysis, no discontinuity could be detected between the sunfish species and other fish. Consequently, we cannot conclude that the centrarchids form a holobaramin, but future research should clarify the baraminic relationships of these fish.

8.4.2.2. Sunflowers.[14] As part of their research into the baraminology of the sunflower family Asteraceae (described in chap. 5), Cavanaugh and Wood performed ANOPA on a morphological data matrix from three different sunflower tribes, Heliantheae, Helenieae, and Eupatorieae. Heliantheae contains the true sunflowers, as well as other common flowers such as zinnias. The precise relationship of these tribes has a contentious history in systematics, with some researchers preferring to place the Helenieae within the Heliantheae and others arguing for their separation as distinct tribes. In 1994, Bremer listed the resolution of the Heliantheae and Helenieae as one of six unresolved problems in the study of Asteraceae evolution.[15]

Both the 1D and 2D ANOPA results revealed two clear groups of organisms, which were subsequently confirmed by the 3D ANOPA distribution. Though well-supported in ANOPA, neither group corresponded to either of the tribes in the data matrix. Instead, both ANOPA groups contained a mingling of taxa from Heliantheae and Helenieae. One of the groups also contained taxa from tribe Eupatorieae, widely considered to be distinct from both Heliantheae and Helenieae. Based on these results, Wood and Cavanaugh concluded that the Heliantheae were continuous with the Helenieae and the Eupatorieae, belonging to a single monobaramin. Because the 3D ANOPA revealed no significant discontinuities, no apobaramins or holobaramins could be identified.

By considering only bifurcating trees, Karis concluded that the Heliantheae evolved from a lineage of Helenieae.[16] Other scientists had previously argued that the Helenieae had a separate evolutionary heritage than the Heliantheae. With the less restrictive perspective provided by 3D ANOPA, Cavanaugh and Wood concluded that some of the problems encountered in classification of these flowers arose from reductionist systematics. By creating separate tribes based on one or two traits, systematists could not successfully describe the classification of these plants.

8.4.3. ANOPA and Trajectories

Occasionally, ANOPA reveals a pattern that has a particularly obvious biological meaning. Specifically, ANOPA can uncover a linear pattern to taxa, wherein the taxa on one end have one phenotype, the taxa on the other have a different phenotype, and the taxa in the middle might be intermediate between the two in some way. Beginning at one end of such a structure, we could follow a gradualistic change from one organism to another. In other cases, extrinsic characteristics (like stratigraphic position for fossil horses) can show a gradation along the ANOPA pattern. Such biologically meaningful patterns are called biological trajectories.[17]

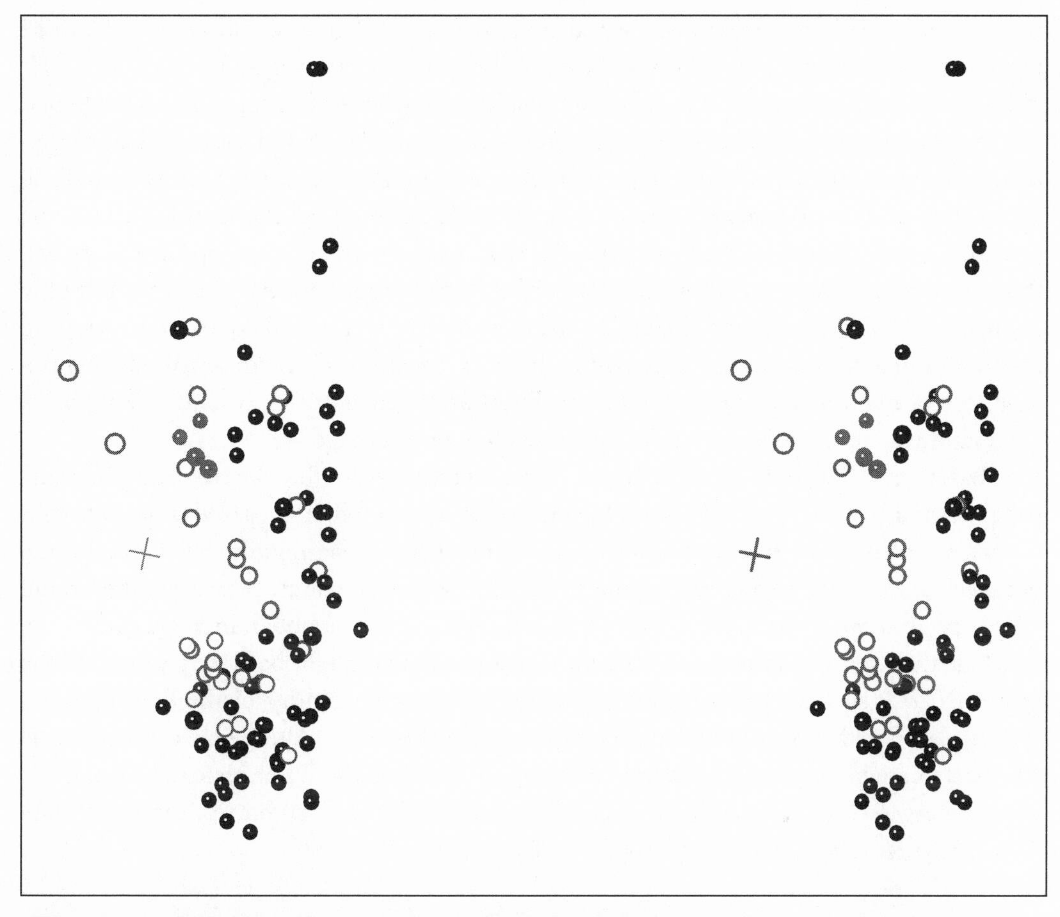

Figure 8.4. A stereo plot of the three-dimensional ANOPA results on a dataset of Heliantheae flowers. Black points represent tribe Heliantheae, white points represent tribe Helenieae, and gray points represent tribe Eupatorieae. The 3D ANOPA disagrees substantially with the cladistic view of the evolution of this family.

8.4.3.1. Horses.[18] As part of a baraminological study of fossil horses, Wood, Cavanaugh, and Wise applied ANOPA to a published dataset of dental, cranial, and post-cranial characters. They discovered that the horse fossils form a true trajectory through the 3D ANOPA space. *Hyracotherium* and *Epihippus* lie at one end of the trajectory, and a large cluster consisting of *Equus, Dinohippus, Neohipparion, Pseudhipparion, Hipparion, Merychippus, Protohippus,* and *Pliohippus* form the other end. Near the end containing Equus, three genera–*Anchitherium, Hypohippus,* and *Megahippus*–conspicuously branch from the trajectory.

By itself, the trajectory suggests that the taxa form a continuous monobaramin, which is confirmed by a baraminic distance analysis. When we look more closely at the Equid 3D ANOPA, we find that the taxa occur along the trajectory in roughly stratigraphic order. Fossils found exclusively in the lowest post-Flood rock layers form one end of the trajectory, while the fossil from the higher levels (including the extant *Equus*) form the other end. More surprising still, we also find that the trajectory closely mirrors the phylogeny of the fossil horses as proposed by MacFadden. Because ANOPA frequently contradicts phylogenetic trees, we must seriously regard this equid trajectory as very good evidence that the fossil equids record an actual path of diversification through character space. If the equid fossils did not represent a diversification series, we would not expect ANOPA to show a trajectory.

8.4.3.2. Flaveria.[19] As described in chapter 5, the genus *Flaveria* consists of plant species that photosynthesize by a C_3 pathway, others that photosynthesize by a C_4 pathway, and still others that are intermediate between the two types. Although it might be tempting to assume that the C_3 species are ancestral to the C_3-C_4, which in turn are ancestral to the C_4 species, this surely cannot be, since all of these species exist in the present. At best, we could only say that the C_3 and C_3-C_4 species could have descended from ancestral populations that were true transitional forms during the evolution of the C_4 species. This predicament highlights a limitation of inferring ancestry. Whereas the linear arrangement of $C_3 \rightarrow C_3$-$C_4 \rightarrow C_4$ has a great intuitive appeal, it is technically not an evolutionary lineage.

Based on a 3D ANOPA of *Flaveria* species, together with other species from the same tribe, we find that the C_3, C_3-C_4, and C_4 species do indeed follow a curved trajectory with C_3 and C_4 plants on either end and C_3-C_4 intermediates in the middle. Furthermore, the non-*Flaveria* C_3 plants also group together with the *Flaveria* C_3 plants, preserving the overall trajectory. In terms of the phylogeny or diversification of this monobaramin, we could only conclude that the C_4 species arose from C_3-C_4 ancestors that might be closely related to the modern C_3-C_4 species. We conclude this because the vast majority of the five thousand species placed in the same monobaramin as *Flaveria* are also C_3.[20] Although we can never be absolutely certain about the ancestral condition of a monobaramin, it seems likely that C_4 plants emerged from C_3 ancestors. Unlike the results with the fossil horses, the *Flaveria* 3D ANOPA pattern did not follow a proposed phylogeny.

8.5. Chapter Summary

The first thing that novice baraminologists notice is that conventional systematics methods do not work for baraminology. All conventional systematics methods concern themselves primarily with identifying the historical path taken by organisms as they evolved. As a result, all conventional systematics methods *assume* that the organisms share a common ancestor.

The baraminic distance and Analysis of Patterns (ANOPA) methods rely on collection of useful baraminological data in the form of a data matrix. Baraminologists may find it useful to examine the attributes of tree-like diversification models for their baramins, but baraminic membership can and should be established independently of such considerations. As much as possible, baraminologists should strive for a balanced and broad-based data matrix. The most useful baraminological data matrices grow from the experience wrought by research with original material.

The baraminic distance method consists of several novel statistics that measure the similarity of organisms based on the percentage of shared characteristics. These distances can then be examined by correlation, which in turn can indicate both significant similarity and significant difference. These similarities and differences can aid in the recognition of continuity or discontinuity.

ANOPA is a multidimensional pattern projection method that treats each character as a separate dimension. Calculations based on the projection of taxa on a multidimensional cylinder can reduce the dimensionality of the data to only three dimensions, allowing viewing of biological character space. Additional methods can reveal the statistical significance of patterns and clusters identified by ANOPA.

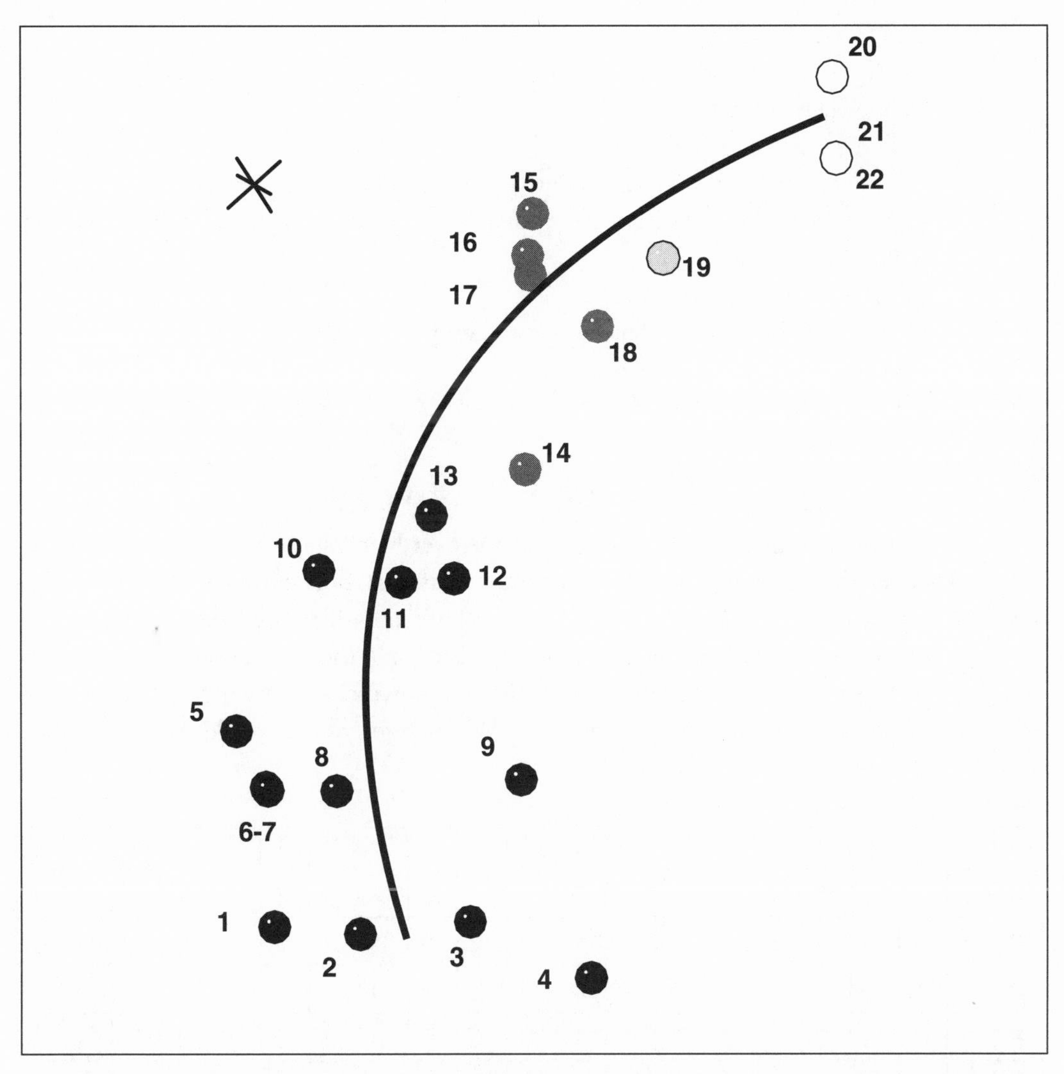

Figure 8.5. *A biological trajectory revealed by three-dimensional ANOPA in subtribe Flaveriinae. Plants that photosynthesize using a C_3 pathway are shown in black, C_4 plants in white, and intermediate plants in gray. Species shown are* Dyssodia paposa *(1),* Tagetes micrantha *(2),* Tagetes lucida *(3),* Flaveria pringlei *(4),* Clappia suaedaefolia *(5),* Chrysactinia mexicana *(6),* Jaumea linearis *(7),* Pseudoclappia arenaria *(8),* Varilla texana *(9),* Haploësthes greggii *(10),* Sartwellia puberula *(11),* Varilla mexicana *(12),* Jaumea carnosa *(13),* Flaveria oppositifolia *(14),* Flaveria linearis *(15),* Flaveria sonorensis *(16),* Flaveria ramosissima *(17),* Flaveria anomala *(18),* Flaveria palmeri *(19),* Flaveria bidentis *(20),* Flaveria bidentis *(21), and* Flaveria trinervia *(22).*

Review Questions

1. Why is holism important to baraminology? What biblical references support your answer?
2. What kind of data should be included in a baraminological data matrix?
3. What kind of data (if any) should be excluded from a baraminological data matrix?
4. Name two internet resources for data matrices.
5. Define baraminic distance.
6. Define character relevance.
7. Define character diversity.
8. Define baraminic signal.
9. How does character relevance relate to calculations of baraminic distance?
10. Explain the baraminic distance correlation test. Justify it using the biological character space concept.
11. What does BDIST calculate?
12. What evidence suggests that the grass family is a holobaramin?
13. Describe Analysis of Patterns.
14. What distinguishes ANOPA from standard phylogenetic methods?

For Further Discussion

1. Is there a perfect data matrix? Could one ever be created? Why or why not?
2. Given the following data matrix, conduct a full baraminic distance analysis, including calculating character relevance, character diversity, baraminic signal, baraminic distances for each pair of organisms, and the correlation of distances for each pair of organisms. Remember that unknown characters are coded as **?** and inapplicable characters are coded as -. What might you conclude about hypothetical species A–J?

A	1	0	0	1	1	?	0	?	1	1	-	0	1	1	?	0	0	1	-	2
B	1	1	1	1	1	1	0	?	0	1	-	0	1	2	1	0	1	1	0	2
C	2	0	0	0	1	1	0	2	1	0	2	0	1	1	1	0	2	1	1	2
D	0	1	0	2	0	2	0	?	1	1	2	0	0	2	0	0	3	?	1	2
E	?	?	1	1	1	2	0	1	1	0	2	0	0	1	?	0	4	?	1	2
F	1	0	1	?	1	2	0	0	1	1	2	0	0	1	1	1	3	?	?	1
G	-	0	0	0	1	2	1	0	1	2	2	0	1	1	1	1	2	0	?	1
H	2	1	0	1	1	0	1	0	1	1	2	0	1	1	1	1	1	0	1	0
I	1	-	1	2	1	-	1	0	-	2	1	0	1	0	?	0	2	0	1	0
J	0	-	1	1	1	1	1	0	-	1	0	0	1	0	-	1	1	1	1	0

SECTION III

Application

CHAPTER 9

Baraminology and Design

9.1. A History of Design

Biological design has become a contentious subject in the past 150 years. Naturally, those opposed to invoking God to explain biological phenomena would argue that design does not explain anything and that we have attained an enlightenment thanks to Darwin's theory of natural selection. Those who favor the inclusion of God in biology would not see Darwin as a bringer of light but perhaps of confusion. As creationists, we invoke God as our ultimate explanation for the origin of life, and we must therefore discuss the implications of baraminology in the area of design theory. Before we do so, however, it would be helpful and informative to review the history of design reasoning in theology and science.

As we know from chapter 1, Aquinas's way of thinking came to dominate the church in the Middle Ages, spawning the so-called "scholastic theology." Scholastics carried the high regard of intellect to an inappropriate extreme, generating what we now call **natural theology**. According to natural theology, God's attributes can be inferred by careful examination of His creation. We eagerly affirm that creation reveals the attributes of the Creator, but the natural theologians distorted the meaning of these passages by vaulting human reason. Although the light of God's truth can be seen in creation, men hate the light because their deeds are evil. As a result, even though God's revelation may be as plain as day, we harbor no delusions that people actually recognize it.

Even today, William Paley's book *Natural Theology* is often considered to be the defining statement of biological design theory. The book describes Paley's famous watch analogy. He argued that a watch with intricate parts all working together toward a common goal gave unmistakable evidence of a watchmaker. By analogy, the intricate parts of organisms all working together toward a common goal (life), suggests that organisms also have a designer. Throughout his book, Paley used various biological phenomena to illustrate theological truths. Near the end of the book, Paley carried natural theology to nearly absurd lengths, claiming that flies, bees, fish, and hares are happy in the role that God had given them.

While natural theology enjoyed its greatest popularity in Britain, other countries had their own ideas about design. In France, Cuvier asserted that organismal design was primarily determined by the environment, or "conditions of existence." Because of the often exquisite balance between biological form and function, Cuvier believed that function determined *all* form. In short, Cuvier equated survival function with design. He rejected any idea that threatened to reduce function to a secondary role in biology, because doing so would reduce the Designer's role in biology. Today, we know this school of biology as **functionalism**.

In Germany, the nature philosophers began to study recurrent themes in anatomy, such as the common structure of vertebrate limbs and skulls. Together with Geoffroy and other philosophical anatomists in France, the German anatomists began to reject the idea of function as the primary cause in biology. They demanded explanations of form for form's sake, explanations that did not invoke function. Philosophical anatomists and nature philosophers came to be called structuralists.

Following the intellectual trend of structuralism, Richard Owen, the first curator of the British Museum (Natural History), attempted to adapt it to a new form of design theory. To Owen, the common plan seen in different organisms represented a concept in the mind of the Creator. Instead of the "conditions of existence," Owen believed that the "unity of plan" testified to God's design. Owen studied comparative anatomy in order to discover the plans used by the Creator to make all life on earth. In his later years, he openly embraced a neoplatonic perspective. Plato believed in two worlds, a world of perfect ideas called "forms" and a world of material reality, referred to as "shadows" of the forms. To Owen, organisms are "shadows" of the divine plan (Plato's "forms").

Much to the chagrin of design enthusiasts, the resolution of this biological disagreement over form and function came not from a Christian biologist seeking the truth of the Creator but from a young naturalist with no particular interest in the Creator. Charles Darwin proposed natural selection as a means of explaining the evidence of functionalism in biology. According to Darwin, normal, random variations and selective breeding could produce great functional perfection. At the same time, Darwin transformed Owen's metaphysical body plan into a physical ancestor. Darwin directly addressed structuralism and functionalism in this extended quote from chapter 6 of *Origin*:

> It is generally acknowledged that all organic beings have been formed on two great laws–Unity of Type, and the Conditions of Existence. By unity of type is meant that fundamental agreement in structure, which we see in organic beings of the same class, and which is quite independent of their habits of life. On my theory, unity of type is explained by unity of descent. The expression of conditions of existence, so often insisted on by the illustrious Cuvier, is fully embraced by the principle of natural selection.

As the vast majority of scientists abandoned design theory, the twentieth century did not see any major innovations in design theory. Most creationists contented themselves with defending Paley and other functionalists. Others have endeavored to make the Creator an anonymous "intelligent designer" rather than the God of the Bible. Most recently, the sophistication of these types of design arguments has improved through the work of William Dembski, Stephen Meyer, Michael Behe, and others.[1]

Qualitatively, each new design argument differs little from Paley's watch. At the heart of all design theories since Paley lies an inference of divine activity from complexity. Although the idea of "divine activity" has become more nebulous over time, the complexity that signals design has become more well-defined. Thaxton used thermodynamics to argue that life itself must be designed.[2] Meyer favors the information content in DNA as a type of designed complexity.[3] In all cases, the design theories take the form of arguments against purely materialistic explanations of origins rather than well-developed scientific models in their own right. As such, these theories might contain value for convincing people of the legitimacy of design, but they do not serve as a basis for building a creation biology.

In the following sections, we will review and discuss several ingredients that must be present in a scientifically useful theory of design. Most importantly, we will develop the impact of baraminology and our refined baramin concept on design. Specifically, how would baraminology change our understanding of God's design? Before developing this new vision of design, we will introduce one of the most misunderstood and contentious concepts in the design debate: mediation.

9.2. Mediation and the Origin of Design

When confronted with a design argument, an ardent evolutionist typically rejects it because it does not explain anything. To an evolutionist, resorting to divine intervention to explain a phenomenon is much worse than merely confessing ignorance of the cause of the phenomenon. Mystified by this attitude, the creationist responds that design is more satisfactory than evolution because of evolution's many unanswered mechanistic questions. To a creationist, merely acknowledging God as Designer and Creator resolves all of these problems. Since neither side seems to understand the other, the battle rages on. If we intend to begin to resolve these questions, we must first understand the queries of both sides.

The evolutionist's primary objection to design targets its lack of discrimination. It is very common for creationists to echo Paley by claiming design for any and every biological phenomenon. The only criterion necessary to invoke design appears to be an awe-inspiring complexity. Evolutionists reject this kind of reasoning. Rather than seek a mechanism to explain biological phenomena, creationists seem merely to surrender to ignorance, chanting, "That's just the way God made it." Evolutionists believe that by invoking design, we creationists discourage further inquiry into the phenomenon.

To be fair to creationists, design encompasses much more than simply declaring "God did it." Although popular creationism has often portrayed design theory in very black and white terms, Marsh's baramins add a deeper dimension to our understanding of God's design. Since Marsh allowed for intrabaraminic diversification, characters specific to certain species of a baramin but lacking in others, no matter how complex, probably arose through historical processes. Accepting historical development of traits does not preclude design, but it does require a different kind of design than direct creation. Under baraminology, God's design must be **mediated**.

When a creationist claims that a biological phenomenon is designed, this need not entail any claim about the *mechanism* of origin. Design could merely describe the intention of the Creator to bring about a certain biological outcome. Thus, the identification of design in terms of God's intentions or plans can be separated from the mechanism of design, the actual mediation by which God brings about his plans. The mechanism of origin by which a design came about could involve any number of mediating events and agents between the design and the Designer.

In the end, identifying some obscure feature of design is infinitely less powerful than explaining the mediation by which an organism came to exist. With a good model of mediated design, creation biologists could actually make specific predictions about what organisms ought to look like. In order to make successful predictions, we must *assume* the plan of design rather than trying to *infer* it. In doing so, we do not assume God's direct action influences every biological event. Instead, we confess that living things exist by His divine intent and sustenance, by His design. His mediated activity in creation can then become a realm of discovery through scientific investigation.

Biblically, we find many references to God's mediating activity. For example, God often sends His revelation through another agent, such as an angel (Dan. 10:4–14) or a prophet (Jon. 1:1–2). Many times, Jesus performed miracles by seemingly unrelated, ritualistic activity. In some cases, the blind receive their sight by Jesus' word alone (Mark 10:46–52). At other times, Jesus touched blind eyes to heal them (Matt. 9:27–31). In at least one situation, Jesus made mud from spit and placed it on a blind man's eyes to restore his vision (Mark 8:22–26). We could list a myriad of other examples, but these suffice to show that mediating divine activity in no way diminishes God's sovereignty or power. Similarly, mediated design does not remove God from His role as Designer and Creator. Considering God's preference for mediated activity in the Scripture, we should expect to find mediated design in creation.

9.3. Designing a New Design

Now we must turn to that daunting task of creating a new design theory, which we cannot do within the limited scope of this chapter. Instead of presenting a new design theory, we will describe what it ought to look like and what it ought to do. By clarifying the nature of design in this way, we hope to give future researchers both a foundation to build on as well as an overall guiding plan. To begin, we introduce two terms to conveniently distinguish between kinds of design. We will use God's *plan* to refer to aspects of design that encompass divine intentionality. God's *implementation* describes the way that God effects his plan in creation, which itself is part of the plan.

We can illustrate these concepts by analogy to building a house. Upon viewing a newly completed home, we might compliment the new homeowner on the clever floor plan that makes particular tasks very convenient. For example, we might be pleased to see a laundry chute that connects the bedrooms on the second floor with the utility room in the basement. Clearly, the history of that laundry chute entails much more than just the desire (design) of the homeowner. It might have begun with the homeowner's request. It was then placed into the technical blueprint by the architect. The home builder observed the specifications for the laundry chute, then ordered the proper materials, and assigned the construction task to one of his laborers. Finally, the interior decorator chose a particular molding to fit around the opening of the chute and a paint color to match the decor of the bedrooms.

All of these stages of its history contributed to the final "design" of the laundry chute. The plan would include everything from the owner's desire to the formal blueprints for the house. The implementation would include the actual construction of the laundry chute.

To construct a new theory of design, we recommend that these two components, plan and implementation, be addressed by recognizing their inherent separateness and interdependence. Illuminating God's plan requires a heavy dose of philosophy and theology before looking to science. The plan cannot be described adequately without involving professionally trained philosophers, biblical scholars, and theologians. Understanding God's implementation requires inquiry into science and should be informed by the activities of those seeking to understand the plan. Since the implementation itself is part of the plan, information about the implementation should inform the understanding of the plan. In this way, as both aspects of design become more sophisticated, they will nurture each other.

9.3.1. The Plan

Because we know that creation is a revelation of God, God's plan of creation must therefore be rooted in His very nature. As creation biologists, we seek to understand God's plan by constructing a model of it. To do so, we need to understand two interwoven areas: theology and biblical studies. We know about God's nature from a careful study of the Bible, because the Bible itself arose from God's nature. We must also have a proper view of the Bible and its inerrancy if we hope to have a proper view of God. Since we have delved into the inerrancy of Scripture in chapter 1, we will not belabor the point here. Instead, we need to emphasize balance in understanding God's nature in order to create an accurate model of design.

In developing a model of God's nature, we must be careful to avoid inserting our own personalities into the model. Unfortunately, people readily adopt a view of God deeply influenced by their own natures or personal histories. For example, a child of authoritarian parents might see God as a harsh judge. If we regard God as *only* a harsh judge, we distort His true nature and stray from a healthy relationship with God. Just as a faulty view of God can harm a person's life, so too a faulty view of God can yield a faulty view of biological design. Paley's emphasis on God's care for His creation led him to conclude that all creatures were happy doing whatever they do.[4] Had he recognized God's holiness and justice, he might have made more sense of organisms that harm other organisms.

In *Natural Theology,* Paley struggled with the existence of poisonous snakes, but he could not make sense of them based on his view of God.[5] Though Paley made some observations, he never confronted the question we all want to know: What kind of God would create such an intricate system for delivering death to other organisms? This question remains to be answered, and that answer will not come from a view of God's nature that overemphasizes His benevolence.

To avoid pitfalls that arise from incomplete knowledge of God or biology, we recommend that our understanding of God's nature and His plan be constructed from two different directions, in a manner analogous to successive approximation. In one "direction," we should continue to expand and refine our view of God, His attributes, and how those attributes contribute to His nature as an integrated whole. This direction should produce general predictions about what the creation ought to look like. These predictions can be tested, allowing for refinement of our model of God's nature. While most of God's attributes are revealed unmistakably in Scripture, our understanding of how those attributes work together leads to applications in biology. As we have seen above, distorted views of God that overemphasize one of His attributes at the expense of the others cannot produce an accurate view of anything else.

In the other "direction," our model of God's plan for creation must necessarily arise from a good understanding of all of biology, as much as humanly possible. By picking and choosing aspects of biology, we might be able to explain them using one or two divine attributes, but we leave much more unexplained in the process. Only by correlating the whole of biology with the whole of God's nature can we explain God's design. Refinement of our model of God's plan in this direction must also proceed by confirmation of predictions, primarily predictions about other undiscovered areas of biology. In the last section of this chapter, we will present a minimal list of biological features that a good design theory should explain.

9.3.2. The Implementation

Despite occasional confessions that organisms have changed drastically since God created them, creationists have not systematized our unique model of historical biology. Consequently, we should at least sympathize with frustrated evolutionists because creationists so often treat design as a declaration rather than a research program. Despite this general disregard for historical biology, numerous creationists have emphasized important events, such as the Fall or Flood, as major influences on the state of extant creatures. Although we discussed biblical history extensively in chapter 3, we hope here to present a limited systematization of important events and other phenomena that influence God's implementation.

The first and most obvious tool that God uses to implement His design is direct creation *ex nihilo* (from nothing). When we think of design, this is the type of design that most often comes to mind. Although skeptics disallow creation, their objections would ultimately rule out historical science in general. For example, we may never understand the warped mind or even identity of Jack the Ripper, but that lack of knowledge in no way diminishes our ability to investigate the historical evidence of the murders attributed to him.

Like Jack the Ripper's reign of terror, direct creation is a historical event, although a singular one. Even though we cannot replicate God's power or action in creation, and we may never fully understand His motivation, the origin of life remains an event that can be investigated by using historical methodology. We can examine the written record of creation in Genesis, we can infer creation possibilities from the nature of God, and we can examine creation itself for clues to understanding the event of its origin.

The remaining methods of implementation mediate divine activity through some kind of created phenomenon. Since the mediators also originate by God's design, we introduce a new level of complexity to design theory by invoking mediation. By acknowledging the reality of mediated design, we can investigate possible mediators by which the design could have come about. Tracing mediated design would ultimately lead to a chain or network of causation grounded in God's *ex nihilo* creation. Although we can be certain that we have barely begun to understand God's mediated design, we need to summarize what we know about mediation here and now in order to plan a fruitful research program for the future.

One of the most obvious mechanisms of mediated design is organismal reproduction. When God created His organisms, He provided them with a mechanism to produce new individuals like themselves. We know that reproduction finds its origin in creation because God told his creatures to be "fruitful, and multiply" (Gen. 1:22–28) and because God created reproductive structures (Gen. 1:11). As reproduction proceeds with its mediating function, God still retains His role as Designer. Job reminds us that God forms our bodies in the womb: "If I did despise the cause of my manservant or of my maidservant, when they contended with me; What then shall I do when God riseth up? . . . Did not he that made me in the womb make him? and did not one fashion us in the womb?" (Job 31:13–15). Reproduction then should not be viewed as God *relinquishing* the power of creation to His creatures but rather as God *sharing* the power of creation with His creatures.

Today, we know that reproduction has instilled within it a great potential for variation. Among sexual eukaryotes, the production of gametes involves a complicated process of chromosomal rearrangement called "crossing over," which results in genetic shuffling of the parental chromosomes. As we now know from studies of bacteria and archaea, these unicellular organisms also undergo a similar process of chromosomal rearrangement, although it occurs somewhat independently of organismal reproduction. These examples illustrate the complex multilayering of God's design. Not only did God endow His creatures with the ability of reproduction, but He designed mechanisms to ensure that reproduction would be variable. As a result, very few organisms are exactly alike.

The geographic pattern of organisms displays another result of God's design, mediated through organismal movement and migration. In Genesis 1, we see a plan evident in the types of organisms created (flying things, swimming things, land creatures, etc.), but the pattern we have today arose after the major geographic upheavals associated with the Flood. The Bible has little to say about animal movement, but one clear passage in Job teaches that God gave the wild ass "the barren land [as] his dwellings" (Job 39:6). Since the asses of Job's time distributed from the ark, God's plan for **biogeography** must have been mediated through the asses' movements and migrations. Similar passages in the same chapter of Job also imply a divine design of animal movement (Job 39:1–5, 26).

Creationists have historically given little attention to the topic of biogeographical design. Instead, we have relegated biogeography to geology. Land bridges, such as the one joining North America to Asia across the present-day Bering Strait, receive the majority of attention, while the actual organismal movement is left to happenstance. We ought not to underestimate the importance of inert, physical factors in biogeography, but we must also recognize tendencies inherent in organisms themselves that implement God's plan for biogeography. These tendencies include non-seasonal migration, seed dispersal, and physiological requirements and preferences. As each of these mechanisms also arose as part of God's plan, we see once again the higher order complexity of design.

Now that we have seen God's plan implemented through both reproduction and dispersal, we also must consider the possibility that God designed organisms to be adaptable over multiple generations, beyond simply reproductive variation. As we have already argued in chapter 2, God probably made living things adaptable to changing environmental conditions. Historically, creationists have attributed these types of changes to degradation of creation, but would it not be more fitting if God created organisms with a special ability to adapt? Considering the special abilities we have already seen in reproduction and dispersal, we must seriously consider the possibility that diversification and adaptation are also a manifestation of design.

As we briefly mentioned above, inert (or abiotic) factors also contribute to the unfolding and maintenance of God's plan. These factors include common events like weather and climate as well as uncommon catastrophes like the Flood. The Bible teaches that the Flood occurred by God's intention, and numerous passages attribute weather to God's control (e.g., Job 28:26; 36:27; Ps. 68:9; 135:7; 147:8). As a result, the development of creation biology relies heavily on the development of creation geology and creation climatology. This reliance goes beyond merely referring to the Flood or the Fall as recorded in the Bible. Unless we can accurately reconstruct the physical history of the planet, we will have difficulty in reconstructing the mediated design that unfolded through biological history.

9.4. Putting It All Together

With this picture of mediated design in mind, we may now turn to aspects of biology that must be incorporated into design theory. Some of these features can be explained by evolution, but others remain enigmatic. A design theory that explains all of them should be preferred to an evolution theory that can explain only a few.

Before we consider anything related directly to biology itself, we must remember that God's creation reveals God's nature. Since revelation is a form of communication, we might find that treating the biological world as a work of literature or art may be more fruitful than a purely scientific approach in designing a design theory. To explain God's design plan, creation biologists should enlist artists, writers, and composers to make sense of the "non-scientific" features of God's design. The biologist may even find his job in constructing a new design theory reduced to clarifying the major patterns in biology for the more artistic minds to explain. We will now briefly list and explain a few of these major patterns that require explanation in design theory.

9.4.1. Biological Similarity

As we have already discussed in chapter 2, organisms do not appear to be designed randomly. Instead, there are an enormous number of similarities that tie organisms together. Evolutionists employ only two terms to describe similarity: **homology** and **analogy**. Homology describes similarities that occur by inheritance from a common ancestor. In contrast, analogous similarities are derived independently, sometimes by chance, sometimes by a common environmental necessity. Ironically, these terms originate in the work of Richard Owen, a biologist who rejected Darwin's theory of evolution.

Unfortunately, creation biologists have not developed an alternative terminology to describe biological similarity. When we consider discontinuity, the broad usage of homology and analogy render them inapplicable to many types of similarity. For example, the similarity of vertebrate limbs (Figure 9.1) are a classic example of homology. Since this similarity extends across groups we know are discontinuous (e.g., birds, land mammals, and whales), it cannot properly be an example of evolutionary homology.

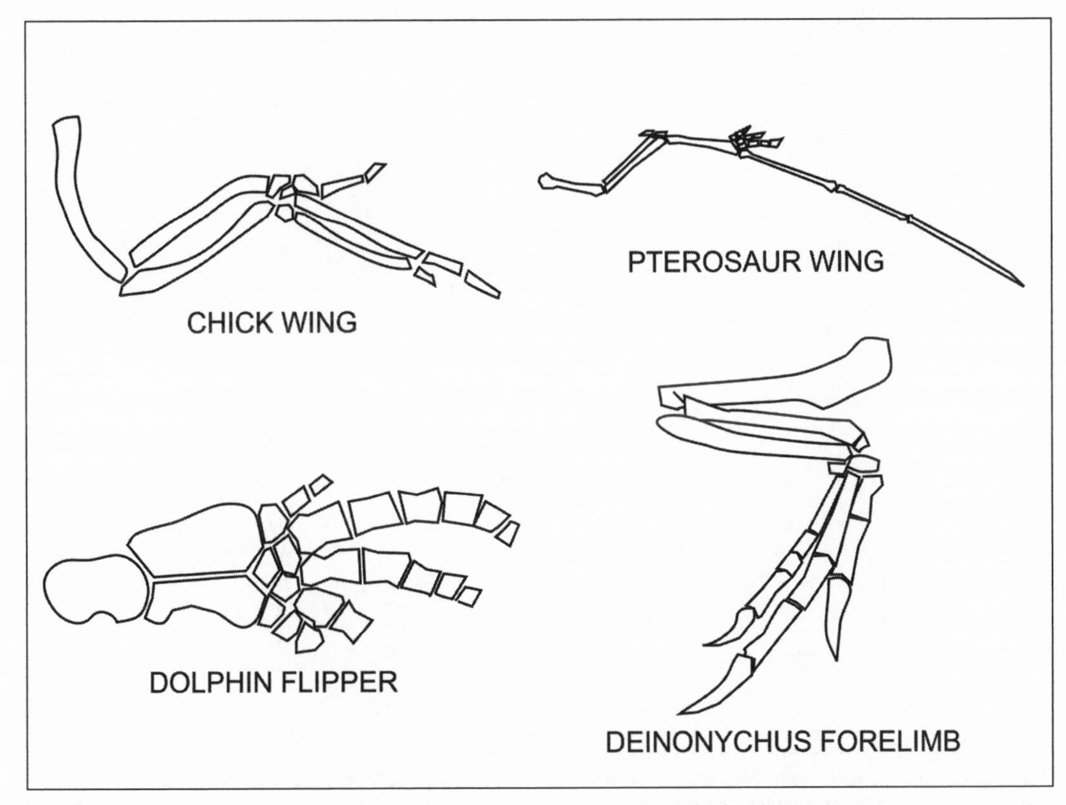

Figure 9.1. *The similarity between vertebrate limbs reveals principles used in the design and creation of organisms. Shown here are the limbs of a chicken, pterosaur, dolphin, and Deinonychus.*

The amazing feature of similarities classified as evolutionary homology is their predictability. If we were to discover a vertebrate never before seen, we could predict details of its skeletal anatomy with startling detail just by virtue of its similarity to other vertebrates. This was put to good use by early structuralists, who successfully predicted numerous anatomical features that functionalists never expected to find.

Even the connotation of *analogy* can be problematic. When we describe these similarities as analogy, biologists immediately think of independent evolution. A famous example of analogy found in most textbooks is the body plan of the shark, dolphin, and ichthyosaur (Figure 9.2), but more specific and surprising homoplasies can be found. Many people know about the sabertooth tigers of western North America, mostly from the celebrated La Brea Tar Pit fossils. Far fewer are aware that the same sabertooth trait also appeared in a group of unrelated South American marsupials called thylacosmilids (Figure 9.3). Evolutionists typically explain these similarities by appealing to some common environmental requirement, such as a streamlined body for swimming.

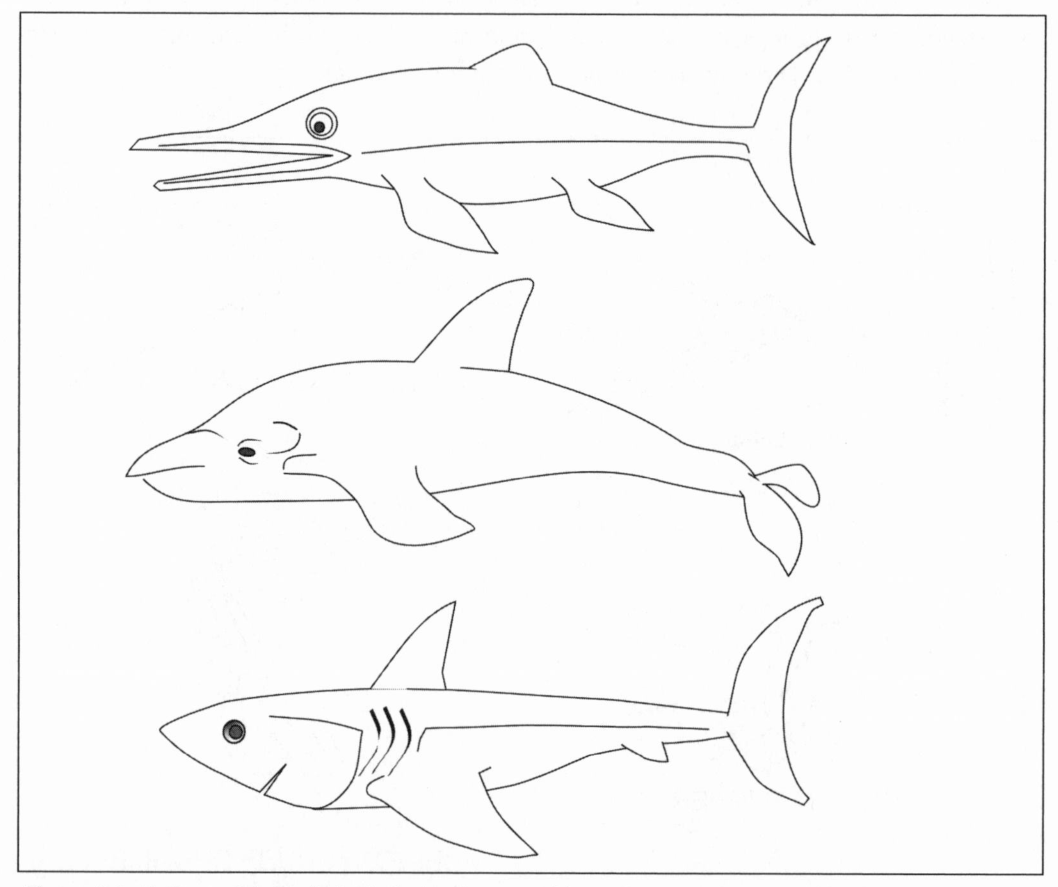

Figure 9.2. *Unlike vertebrate limb similarity, the resemblance between the body plans of the* Ichthyosaur, *dolphin, and shark are superficial.*

Because of the confusion of terminology, past creationist explanations of biological similarity are unsatisfactory. Creationists usually explain "homology" much as Owen did: God merely reused the same basic body plan in different organisms. Creationists have hesitated to venture beyond the assertion of a common designer. Marsh actually rejected such "speculation" as irreverent and impossible. In *Evolution, Creation, and Science,* Marsh wrote:[6]

> The evolutionist [asks] another question; ". . . What possible reason would a Creator have for forming them in such a way that the embryonic man and the adult sea squirt would both have notochords?" To this the creationist replies, "Who is man to attempt to assume why the Supreme Intelligence did this or that?" The evolutionist's question is absurd.

Marsh makes a subtle error in this objection to the question of notochord homology posed by his hypothetical evolutionist. He would be correct if the "Supreme Intelligence"

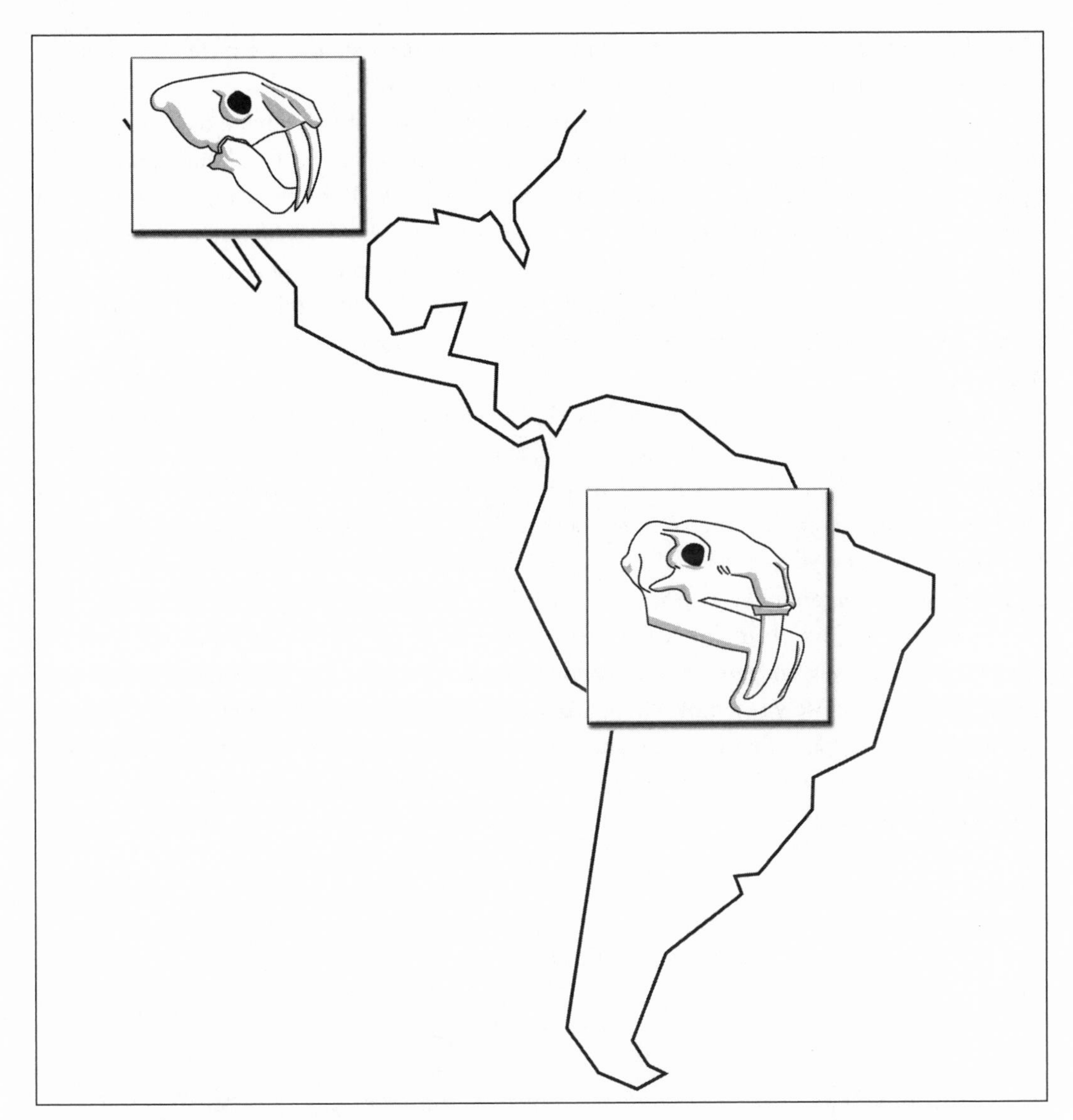

Figure 9.3. *Enlarged canines (the sabertooth trait) emerged in several different baramins after the Flood. Illustrated here are the North American sabertooth tiger* Smilodon *and the South American sabertooth marsupial* Thylacosmilus.

was fundamentally inscrutable and unknowable; however, as we argued in chapters 1 and 2, the Bible depicts a God who wants to be known by His people. Marsh's objection may have repelled the argument of his hypothetical evolutionist, but in the end, stubborn refusal to address the question of biological similarity only leads to defeat. Creationists must attempt to explain, in detail, the mechanism and meaning of similarity.

Under baraminology, we find a number of kinds of similarity, each of which probably comes from different aspects of the design plan or implementation. Some similarity will occur simply because of the fidelity of reproduction. Strictly speaking, this would be a kind of evolutionary homology. Other "homologous" similarities, such as vertebrate limbs, undoubtedly arise as part of the design plan. These similarities require a different kind of explanation, keeping in mind the predictability of these similarities.

Some analogy also comes from the design plan, but other "analogous" similarities appear to be historical occurrences. For example, the body plan of dolphins and sharks appears to be a planned feature, since all dolphins and sharks have common body plans. In contrast, the sabertooth trait emerged several times during the diversification of carnivores after the Flood. While sabertooth similarities are analogous in the sense of "independently derived," the dolphin/shark body plan and the sabertooth trait are independently derived from different sources. The post-Flood emergence of the sabertooth trait indicates a mediated design.

Evidence recently uncovered implies that analogous-type similarity can occasionally arise within a baramin. For example, some salamander species from the genus *Bolitoglossa* have webbed feet. The anatomical similarity between these species is superficial, however, as different species accomplish the webbed-footedness with slightly different anatomical modifications of the toes (Figure 9.4). More research in this area is necessary, but it appears that similarity within baramins does not always arise by common ancestry.

Studies of biological similarity between organisms from different baramins may help to explain one of the most nagging questions in creationism: the similarity of humans and chimpanzees. Humans closely resemble chimpanzees in both anatomy and genetics, leading some evolutionists to propose that humans are nothing more than a third species of chimpanzee. Similarities of this kind will find their explanation in God's design plan, rather than in a particular implementation.

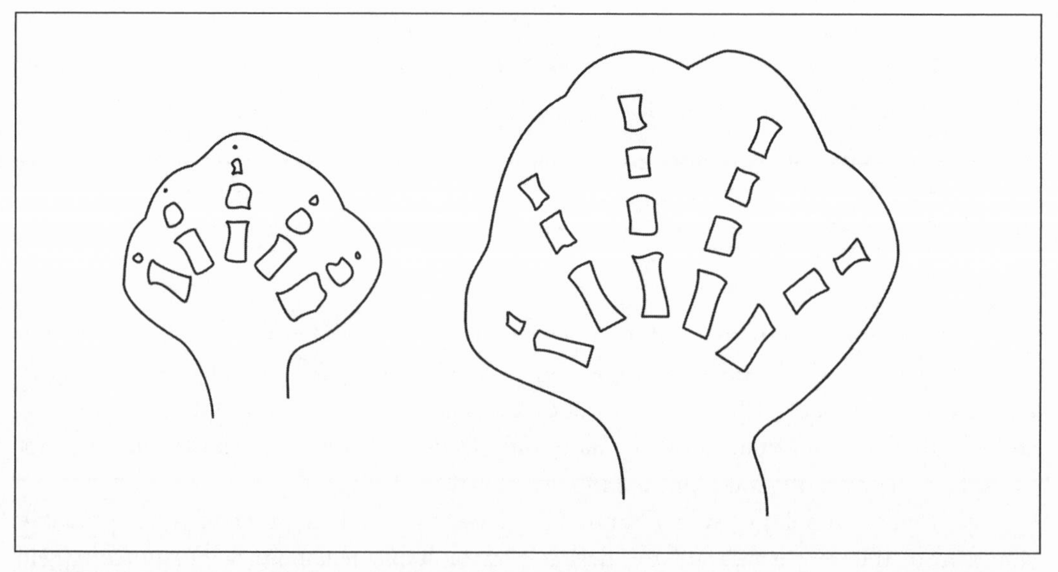

Figure 9.4. *Even within a baramin, similar traits may emerge independently. Here we see the different anatomical changes that produced webbed feet in two salamander species of the genus* Bolitoglossa.

9.4.2. *The Pattern of Life*

Since we discussed this feature of biology extensively in chapter 2, we will not belabor the point here. We should remember, though, that baraminology has the potential to explain more aspects of the pattern than evolution, which is wedded to a single pattern. The hierarchical pattern of Owen, the linear pattern of Aristotle, and even more complex patterns can be accommodated in baraminology. Our understanding of the pattern of life will mature and grow as we investigate more biological similarities.

9.4.3. *Functional Adaptation*

Since adaptation is central to evolutionary research, we need not repeat examples of adaptation here. Instead, we reiterate that functional adaptation does not necessitate the conclusion of unmediated design. When we find a startling functional adaptation within a baramin, we first must determine if the adaptation is shared by all species of the baramin. If it is, then we might reasonably conclude that it arose by unmediated design. If there are one or two species of the baramin who lack the adaptation, we might still conclude that the adaptation arose by unmediated design but that it has subsequently been lost in the few species that lack it. If only one or two species of the baramin possess the adaptation, the conclusion of mediated design is probably warranted. A good design theory must differentiate and explain unmediated and mediated functional design.

As we study functional design, we must always keep in mind post-Flood diversification. Much evidence indicates that the period following the Flood was one of explosive speciation, termed *diversification* (see chap. 11). Since many of these changes altered the physiology and life habits of the newly arising species, diversification should play a large role in the development of a theory of functional design. As in all mediated design, we must remember that the mechanism of diversification was itself designed by God. Consequently, we should not limit our theorizing on the mechanism of diversification to natural selection on allelic variations. We must remain open to special mechanisms that God created specifically to induce diversification and adaptation.

9.4.4. *Biological Imperfection*

A common criticism of the creation model is the existence (some would say the ubiquity) of imperfections. Surely, the skeptic says, a creator could have done a better job than this! Imperfections can include such things as predation and disease, apparently useless features like "junk" DNA, and odd or "improvised" structures like the panda's thumb. Of all the mysteries of design theory, creationists have tackled this one with the most vigor, primarily because of the vigorous attacks of skeptics. Although we will deal with the problem of imperfection in more detail in the following chapter, we need to mention it here because it has a great influence on our understanding and explanation of design.

Many imperfections exhibit a function that harms other organisms. Pathogens, predators, and parasites all fall in this category of imperfection. Creationists attribute such "natural evil" to the Fall but also offer a panoply of mechanisms to account for specific features of modern imperfection. Most creationists reject the idea that God created anything *ex nihilo* after Creation Week, and consequently they seek to explain imperfection as a change of what already existed. The most popular theory proposes that much of the imperfection arose because of some kind of degradation that began at the Fall. Everything from pathogenicity to predation has been attributed to this theory.

9.4.5. Non-Survival Features

Often ignored by creationists, features of living things unrelated to their survival ought to be incorporated into the design theory to increase its explanatory power. For example, living things often possess a beauty superfluous to the requirements of existence. This is particularly true of organisms found at the bottom of the oceans. Despite the complete lack of sunlight at those depths, the organisms living there often display a startling beauty far beyond any possible need. Since creation is a revelation of God, His nature is our basis for understanding the beauty of living things. More aesthetic studies of biological phenomena will aid the incorporation of beauty into design theory.

One attribute of God's character–abundance–can provide a basis for understanding these strange non-survival characteristics. The strict functionalist paradigm fails most spectacularly when faced with the amazing number of living things. If God's primary design goal was to produce an enduring habitation for man, why did He create so many millions of species? Surely He could have produced a far simpler and more efficient creation with perhaps a hundred species or baramins. Although this question confounds the functionalist, we may explain the problem very simply as a manifestation of God's abundance. The Bible reveals a God who generously and gladly provides above and beyond anything anyone

Figure 9.5. *Beauty is a ubiquitous part of God's design plan for creation. Even though no light penetrates to the depths of the ocean, the organisms that live there, such as these tube worms, exhibit a surprising beauty.*

could ever need. Such a God would happily create millions of species precisely because humans do not *need* them.

This area of biology needs further attention. Beyond anything else, non-survival features of biology have the potential to set design theory apart from other theories of the origin of life. Evolution offers few explanations for the origin and preservation of features that do not directly benefit the organism's survival. Within creationism, therefore, not only do we have a basis to explain data recognized by all, but we also have the ability to observe data that only a design theory would predict.

9.4.6. Macrobiogeography

We have devoted the final chapter of this textbook to the subject of biogeography, but we mention it here because it is a very neglected field in creation biology, particularly with respect to design. Because of the vast multitude of mediators that work together to produce the modern distribution of organisms, biogeography may prove to be the most difficult area of biology to subsume within a comprehensive theory of design. As we noted above, geographic distribution can be influenced by extrinsic factors such as climate or barriers and by intrinsic factors such as physiology and mobility.

Within the creation model, we must recognize two distinct biogeographic periods with a period of transition (the Flood) between them. Because of its relatively undisturbed proximity to the original creation, the biogeography of the pre-Flood world likely depends greatly on God's original plan of design and unmediated implementation of that plan during Creation Week. To truly understand biogeography before the Flood, we first need a reliable estimation of the coastline of the pre-Flood land masses. Much more work in this area remains to be done. One important idea first proposed by Harold Clark is that the geography of the pre-Flood world and the distribution of organisms can be used to explain the fossil record.[7] If correct, the distribution of organisms in the earth's crust may aid us in understanding what the pre-Flood biogeography was like.

The Flood completely reworked the land at the same time that it altered the history of living organisms. While the organisms diversified and adapted to the post-Flood world, they also dispersed from **centers of survival**. For terrestrial animal baramins, Ararat is the center of survival, but for other baramins, centers of survival could exist anywhere on the earth and any one baramin might have more than one center of survival. Post-Flood biogeography will depend on theories of diversification and adaptation, post-Flood geological and climatological recovery, post-Flood paleontology, and basic baraminology. Creationists have rarely considered biogeography in any other context than a response to critics, but by framing biogeography in a larger design theory, we should be able to facilitate future research. As we will explain in the final chapter of this book, creationists have a powerful basis for understanding modern biogeography.

9.5. Chapter Summary

Much to the chagrin of design enthusiasts, the resolution of the form/function argument came from a young naturalist with no particular interest in the Creator. Charles Darwin proposed natural selection as a means of explaining the evidence of functionalism in biology. According to Darwin, normal, random variations and selective breeding could produce great functional perfection. At the same time, Darwin transformed Owen's metaphysical, structuralist body plan into a physical ancestor.

The evolutionist's primary objection to design targets its lack of discrimination. It is very common for creationists to claim design for any and every biological phenomenon; however, design encompasses much more than simply declaring "God did it." Under baraminology, God's design must be mediated. With a good model of mediated design, creation biologists could actually make specific predictions about what organisms ought to look like.

To construct a new theory of design, it is recommended that two components, plan and implementation, be put to use. God's plan of creation must therefore be rooted in His very nature. The design implementation refers to the methods and techniques that God used to accomplish His plan. The first and most obvious tool that God used to implement His design was direct creation *ex nihilo* (from nothing). When we think of design, this is the type of design that most often comes to mind. The remaining methods of implementation mediate divine activity through some kind of created phenomenon (e.g., migration or reproduction).

Review Questions

1. Define natural theology.
2. Define functionalism.
3. Name two structuralists.
4. What significant role, if any, did "nature philosophers" play in the development of the science of anatomy?
5. Define mediated design and defend it biblically.
6. Use the house-building analogy to explain design.
7. If baraminology is correct, why must design be mediated?
8. List six features of biology that design should explain.
9. What are centers of survival? How do they relate to the design of biogeography?
10. Design can be divided into two helpful concepts. Name them.
11. What did Paley believe about living organisms and their response to God's design?
12. How could we use a "successive approximation" strategy to understand the nature of God?
13. How does historical biology help us to understand design?
14. Why is biological similarity so important to design theory?
15. How did Marsh err in his understanding of notochord homology?
16. How did Marsh's baraminology add depth to design theory?
17. List the attributes of creation that design should explain.
18. How are the design plan and implementation interconnected?
19. Give biblical references that justify mediated design.

For Further Discussion

1. How has William Paley's *Natural Theology* influenced modern conceptions and misconceptions about design?
2. Why is a new design theory needed?
3. Should there be one design theory or many? Why?
4. Explain what you perceive to be the most important components of a biologically relevant design theory.
5. Is it possible to model God's design plan without scientific information? Why or why not?
6. What future events might change our current view of design and baraminology? Justify your answer biblically.
7. As organisms adapt to environmental conditions, could they become less similar to members of their own baramin and more similar to members of a different baramin? Is there a "core" set of similarities to other members of their baramin that never change?
8. Has God's design plan changed in response to major global events such as the Fall or the Flood?
9. How can we trace mediated design?
10. Should design theory be entirely scientific? Why or why not?

CHAPTER 10
Biological Imperfection

10.1. Introduction

The sundry imperfections detectable in the living world have served as the focal point of many an argument between design advocates and skeptics. This is by necessity closely related to the question of design. We might summarize the question of imperfection like this: "If God is a perfect designer, why is His design not perfect?" Examples of imperfection include vestigial organs, junk DNA, and disease. Most readers will recognize examples of "natural evil" as common imperfections. As the Book of Job testifies, man has long struggled with reconciling a benevolent, omnipotent God with the existence of evil. In this chapter, we do not intend to try to resolve this ancient problem. Instead, we hope to clarify the question specifically as it applies to biology

We prefer the term **biological imperfection** rather than **natural evil**. Evil is a moral condition found in moral creatures. Since the vast majority of God's creatures are not moral creatures, they cannot be evil in the strict sense. Although non-moral objects may participate in evil as devices of moral creatures, we do not know for certain the true nature of natural evil. It would be hasty to conclude that all biological "natural evil" is the device of an evil agent. To distance ourselves from the improper attribution of evil to creatures that do not deserve the attribution, we recommend the use of the term *biological imperfection*.

Within the category of imperfection, we place biological phenomena that appear to be contrary to the nature of God the Designer. Naturally, this brings us to our first problem: How could we possibly know when something is contrary to the nature of God? What we classify as biological imperfection is actually whatever runs contrary to our *perception* of the nature of God. God's nature and our perception of it can be very different. As a result, creation biologists must exercise great care and caution when classifying biological phenomena as imperfection.

The late Stephen Jay Gould illustrated this problem of perception very well. In his famous essay "The Panda's Thumb," he began by discussing Darwin's follow up to *Origin*, published in 1862: *On the Various Contrivances by Which British and Foreign Orchids Are Fertilized by Insects*.[1] After describing the various means by which orchid plants lure pollinating insects to their flowers, Gould noted that in every case, the structures used are modifications of flower parts found in all flowering plants. He then made the leap to judging the Designer:[2] "If God had designed a beautiful machine to reflect his wisdom and power, surely he would not have used a collection of parts generally fashioned for other purposes." Gould assumed that he knew what motivates God to design and create, but any careful reader might wonder *how* Gould knew God so well that he could say with certainty what God would and would not do.

Figure 10.1. *Orchids like this Pink Lady's Slipper (*Cypripedium acaule*) exhibit a wide variety of pollination mechanisms, all of which derive from a basic set of floral parts common to many different flowers.*

Instead of besmirching a "straw-man" designer, Gould might have simply made the point that the combination of natural selection and common ancestry explains orchid anatomy very well. Thus, those who maintain that the orchid anatomy came from the mind and will of God must explain why the flowers appear to be naturally "contrived." Gould's perception of God falls far short of reality, but his mistake serves to warn us of the same kind of hubris. As we approach the issue of imperfection, we must be careful not to impute to God any motivations or values that He does not possess. The best place to begin our study of imperfection, therefore, is God's revelation to us in the Bible.

10.2. What Precisely Is Imperfect?

"For we know that the whole creation groaneth and travaileth in pain together until now" (Rom. 8:22). We begin our study of imperfection with the groaning of creation, which stands as a vocal testimony against theistic evolutionists who would have us believe that evolution, with its death and suffering, was God's method of creation. As we shall see in this survey, the Bible clearly teaches that something is wrong with God's creation. It no longer occupies that "very good" state in which God created it.

From Genesis 3, we know that the problem with creation began with Adam and Eve's sin. Because we have already covered the history of the Curse in great detail in chapter 2, we now turn specifically to the effects of the Curse. Traditionally, creationists have attributed every conceivable "evil" to the Curse, from human and animal death to poisons, toxins, and pathogenic bacteria.[3] What biblical evidence can we use to support such claims? The two passages most pertinent to this question are found in Genesis 3 and Isaiah 11. Genesis 3 records the actual Curse, while Isaiah 11 records a vision of the future in which the effects of the Curse have been lifted.

In Genesis 3, God curses three individuals: the snake, Eve, and Adam. The snake is cursed *above* all other animals, from which we infer that all other animals were also included in the Curse, though their specific curses were not recorded. The snake was doomed to crawl on its belly and to eat dust. Eve was cursed with sorrow in childbirth and conflict in marriage. God cursed the ground with weeds for Adam's sake, and clearly laid out the promise of physical death by saying, "For dust thou art, and unto dust shalt thou return" (Gen. 3:19). Young-earth creationists have been criticized for inferring from this passage that animals did not die before the Fall, since no curse of death on the animals appears in Genesis 3. Interestingly, no curse of death on Eve appears in Genesis 3 either; thus, it is reasonable to conclude that not every effect of the Curse receives an explicit statement.

To infer more effects of the Curse, we turn to Isaiah 11. Isaiah describes in his vision the future coming of the Branch of Jesse. At this time, "The wolf will dwell with the lamb, And the leopard will lie down with the kid, And the calf and the young lion and the fatling together; And a little boy will lead them. Also the cow and the bear will graze, Their young will lie down together; And the lion will eat straw like the ox. And the nursing child will play by the hole of the cobra, And the weaned child will put his hand on the viper's den" (Isa. 11:6–8 NASB).

The herbivory of the bear and lion described in this passage hearkens back to the perfect condition in which the animals were created, "And to every beast of the earth . . . I have given every green herb for meat" (Gen. 1:30). Taken together, Genesis 1:30 and Isaiah 11:6–8 strongly imply that carnivory was unknown before the Fall, which would be consistent with God's curse of the snake "above" the other animals implying animal death as part of the curse (Gen. 3:14). The description of the harmless venomous snakes in Isaiah 11:8 also implies that these organisms were not the dangerous creatures they are today.

By ruling out animal death and carnivory, we may also rule out other aspects of our modern biological world. Since pathogenic microorganisms often cause death, we may infer that pathogenicity and parasitism did not exist prior to the Fall. Organisms that possess structures that appear to be "designed to kill," such as the venom-injecting fangs of the viper, also must have existed in some different, harmless state before the Curse. Finally, we might also attribute some general degeneration to the Fall. Since Adam and Eve would have lived forever with no death prior to their sin, presumably they would not have aged as we do today. It seems reasonable to infer then that the degeneration of aging also arose by the Curse.

As we study the biblical record, we must be careful not to overstep the bounds of what Scripture actually teaches about the Fall. For example, we could easily say that there were no parasites before the Fall, but that statement could be misleading. While it is true that no organisms would have been parasitic before the Fall, the ancestors of our modern parasites probably existed before the Fall in some kind of non-parasitic lifestyle. We must carefully differentiate precisely what imperfection arose at the Fall, lest we blame the Fall for many of God's good creations.

Keeping in mind the importance of recognizing the precise nature of imperfection, we can categorize phenomena related to the Curse as we examine the living world. First, we find some imperfections that relate primarily to the behavior or ecology of organisms, in the broadest sense. For example, what may be a lovely wildflower in the forest becomes a nuisance when choking out the tomato plants in our gardens. The plant becomes imperfect by where it grows, not because of intrinsic properties of the plant. In the same sense, the lion becomes imperfect by eating the young goat, not by anything within the lion itself. Even the anthrax bacterium only kills when inhaled or ingested by a susceptible host.

Beyond simple behavioral or ecological imperfection, we also find a more challenging form of imperfection in structures or substances that seem to be designed to harm other organisms. Examples include the complex multiprotein toxins produced by some bacteria, the poisonous substances produced by some plants, and the amazing venom-injection system of snakes. While we might easily imagine a wildflower becoming imperfect by growing where it should not, the origin of complex toxins and other imperfect structures will require a more sophisticated theory of design and imperfection.

Finally, we find a third category of imperfections that arise because of degeneration from the originally perfect creation. This degeneration could manifest itself as flaws in creation or even in the design mediation (such as the diversification mechanism). Examples of degeneration include mutations that lead to such genetic diseases as cystic fibrosis. Unlike the previous categories, this group of imperfection-related features is tied together by the mechanism of origin, rather than any description of the symptoms. For example, it might turn out that weeds (a "behavioral" imperfection) are a degeneration of the pre-Fall ecological design.

10.3. Imperfection in Biology

Within the theoretical framework of baraminology, we find a bevy of new tools to help explain the existence of imperfection. In contrast, rigid species fixity is much less explanatory. If species cannot change, any imperfections exhibited by species must necessarily trace their origin back to the ancestral members of the species that were created or cursed by God. Within baraminology, we are not so limited, and we may propose that imperfections arose during the history of baramins. We may propose and investigate any number of historical mechanisms to explain modern imperfection within the creation biology model, providing the basis for future scientific research. Such an approach can lead to unexpected conclusions.

In addition to the mechanism by which imperfections arose, the rate at which they appear should also inform our imperfection theories. If we maintain that some imperfections arose by degeneration from a perfected state, we might expect that these imperfections would originate gradually throughout history. Even today, we should expect to see new degenerative imperfections emerging. At the other extreme, we should not rule out the possibility that some imperfections emanated immediately from the Curse. Certainly, weeds and animal death would fit into this category of imperfection. More research is necessary to illuminate the relationship between the slowly appearing, degenerative imperfections and immediately appearing, Curse-related imperfections.

In this section, we will present a number of examples that illustrate the importance of studying both the mechanism and rate of origin of imperfections. The first four examples cover issues of mechanism, and the last two deal with the rate of imperfection. We will begin with structures that are apparently useless followed by claims of poor design. We will conclude this section with examples of elegantly designed biological phenomena that appear to be designed to harm other creatures. All of these examples will illustrate the importance of applying baraminology to problems of design and imperfection.

10.3.1. Mechanism: Vestigial organs

As we mentioned above, one mechanism to explain imperfection is degeneration. In biology, degeneration could manifest itself in a variety of ways, but most are identified usually by comparison to other organisms. Organs that appear robust and useful in one organism might appear reduced to near uselessness in another. These organs have come to be called "vestigial" organs. In some cases, such as eyeless cave fish, we can actually find evidence for the degeneration itself. In the perpetual darkness of caves, population of fish that wander in often have substantially reduced eyes when compared to closely-related fish on the surface. In the case of *Astyanax mexicanus,* the eyeless fish belong to the same species as normal fish on the surface.[4] This close relationship illustrates the significant continuity between the fish and therefore places them in the same monobaramin. Since they share such close continuity, differences between the eyeless and normal fish must have arisen during history rather than from unmediated activity of God at creation. Consequently, it would be improper to conclude that God created eyeless cave fish to live in caves.

Despite the explanatory power of the baraminology approach, creationists have reacted to vestigial organs in an extremely inconsistent manner. Instead of relying on careful baraminology studies, most creationists demand that vestigial organs cannot be vestigial if they have any function at all. At the same time, when confronted with the vestigial eyes of cave-dwelling animals, creationists readily accept that degeneration produces such changes in normally-sighted species. Certainly, if degeneration can produce vestigial eyes in cave animals, then it could also produce vestigial organs in the human body. There is no logical reason to accept degeneration in one case but reject it in another.

Similar to the question of vestigial organs, "junk" DNA also provides an example of apparently useless structures that admits a different explanation. While some junk DNA may be truly vestigial, the study of the complex genome has only just begun. Because we currently know very little about the function of DNA sequences in the genome, it would be very presumptuous of us to claim that a particular sequence has no function in modern organisms. Overall, however, there does appear to be a certain validity to the junk DNA claim. Since functional human genes occupy less than 1 percent of the sequence of the human genome,[5] we find it very hard to believe that the *all* of the remaining 99 percent of the genome performs a vital function in the human cell. It could be that some (perhaps much) of the non-gene DNA in the human genome arose by degeneration from previously functional sequences.

10.3.2. Mechanism: The Vertebrate Retina

A second example of imperfection requires a more broad explanation. In some cases, claims of vestigial or maladapted structures appear in all members of a baramin or even in many baramins. Obviously in such cases, we have no basis for inferring that such organs or structures are truly vestigial. Instead of being imperfections, cases such as these most likely trace their origin back to God's plan for creation. As a result, we must consider these types of structures as questions of design. As we do so, we must not constrain ourselves by assuming that God always designs perfectly engineered structures. In some cases, He could value aesthetics or revelation over efficiency. We must take this truth into consideration as we seek to explain seemingly "poorly engineered" structures.

Judging by the number of apologetic articles written on the subject, the anatomy of the vertebrate retina is a favorite example of "poor engineering."[6] The argument for the imperfection of the vertebrate retina centers on the fact that the nerves that connect each retinal cell to the brain run over the surface of the light-sensing cells, then through a hole in the retina on their way to the brain. As a result of this arrangement, light entering the eye must pass through the nerves as they approach the photoreceptors. The hole in the retina results in a "blind spot" in the visual field because of the lack of photoreceptors. In contrast, the eyes of cephalopods have retinas with nerve-connections facing the back of the eye, necessitating neither the obscuring of the retina by the nerves nor the blind spot.

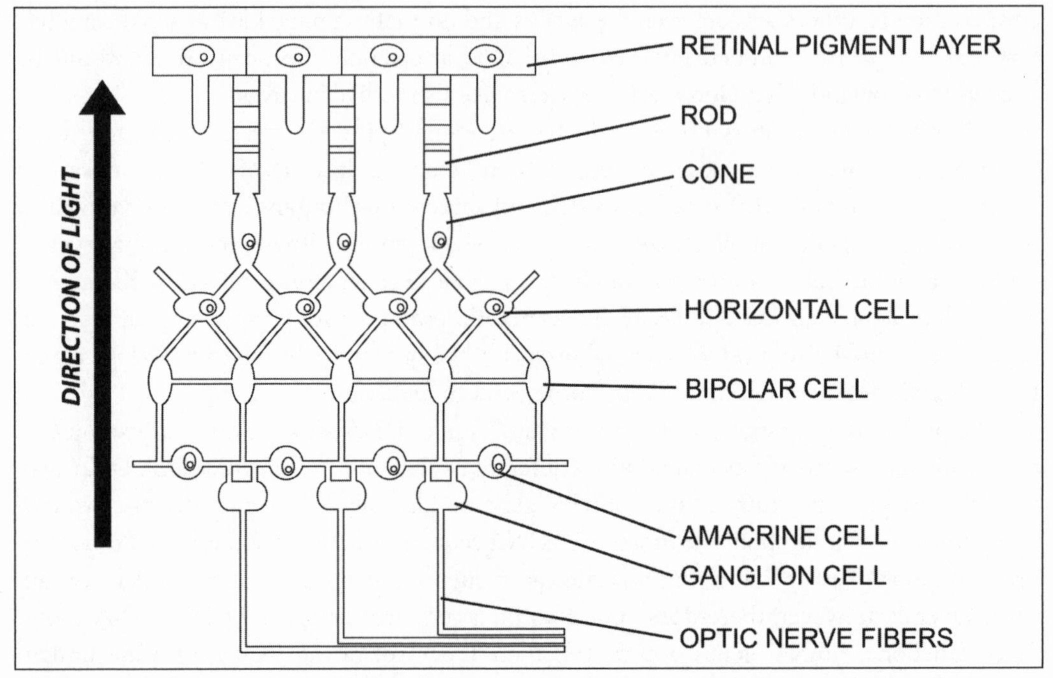

Figure 10.2. *Unlike the eyes of squid and octopi, the nerves of the vertebrate eyes run along the surface of the retina. As a result, light coming into the eye must first pass through a layer of nerves before stimulating the light receptors of the retina. Since all vertebrates possess this attribute, we may attribute it to God's design plan.*

Richard Dawkins makes the best argument for the imperfection of the vertebrate retina by leaving it entirely intuitive. Indeed, he admits that the nerves probably do not inhibit light perception to any significant degree; nevertheless, the arrangement of the vertebrate retina does run counter to the instinctive plan that a human engineer might propose.[7] We could respond to this criticism by demonstrating the functional importance of the inverted retina, but that would miss the point of the problem. Given that God designed the world, and given that He is omniscient and omnipotent, He *could* have designed the retina with the nerves running behind it, since that is precisely the design He chose for the cephalopod.

Even still, demonstration of the exquisite design of the eye does one good thing for creationists: It places the inverted retina problem into the proper field of inquiry: design. Since all vertebrates have inverted retinas, we probably cannot explain its presence as a result of the Fall. Thus, it becomes an issue of either design or imperfection, and since vertebrate retinas are demonstrably well-designed functionally speaking (since we can see), it seems a stretch to classify the vertebrate retina as an imperfection. Instead of viewing it as an imperfection, creationists might ask why God chose the inverted retina for vertebrates but not for cephalopods. To answer this question, we might turn to functionalism or a more holistic design model as described in the previous chapter.

10.3.3. Mechanism: Carnivorous baramins

Among the more challenging imperfection phenomena to explain, some baramins share the same Curse-related imperfection among all of their members. All snakes, for example, are completely carnivorous and display no propensity for eating vegetation, as creationists insist they must have done prior to the Fall. Carnivory actually raises two separate problems for the creationist. First, how did organisms become so apparently well-designed for carnivory? Second, how could creatures so well-designed for carnivory survive as obligate herbivores?

Generally, creationists have been very hesitant to admit that organisms which *appear* to be designed to hunt, kill, and eat meat *were* designed to hunt, kill, and eat meat. Creationists frequently propose that carnivory developed gradually through various stages involving scavenging and increasing aggression. The motivation for this explanation appears to be a reluctance to attribute to a benevolent Creator the well-designed anatomy and behavior associated with carnivory.

We must admit, however, that it is possible that God redesigned several baramins for carnivory at the Fall. In a world of animal sickness and death, mechanisms would be required to control population sizes, eliminate sick organisms to keep populations healthy, and remove rotting carcasses. In such a world, predators would provide a beneficial function to the persistence of the revelation. Thus, creationists should not be so hesitant in marveling at the good designs associated with hunting and killing. Instead of attributing these designs to the original creation, we may associate them with the "redesign" that occurred at the Fall.

As for living carnivores eating vegetation, much evidence of carnivores eating plant material can be cited. When viewing the dietary habits of organisms, we actually find that carnivorous baramins form a continuum, with strict carnivores at one end and omnivores at the other. Since a continuous pattern of carnivory might be expected from a group of created herbivores becoming carnivores, this overall pattern provides support for the creation model.

The various mammal families in the order Carnivora illustrate the dietary continuum. Some of the families, particularly Felidae, exclusively eat meat, but most of the families are not obligate carnivores. The raccoons and bears, for example, will eat whatever is available, including human refuse. At least one member of the bear family, the giant panda, is an obligate herbivore. Although no creationist has evaluated the baraminic position of the giant panda, it does offer some hope for understanding how apparent carnivores could have survived as strict herbivores before the Fall. The very existence of the herbivorous panda validates the possibility that other carnivores could have previously been herbivores, both behaviorally and dietarily.

Unlike other members of the order Carnivora, the giant panda has several features that allow it to tolerate the bamboo shoots that comprise the entirety of its diet. The giant panda's jaw has enlarged muscles for grinding the bamboo, and its esophagus has an inner lining that protects it from damage from sharp bamboo fragments. Most notoriously, the radial sesamoid bone of the panda's forepaws are enlarged into a thumb-like appendage for stripping leaves from bamboo shoots. Because the giant panda does not digest bamboo well, it must eat inordinate amounts of the plant in order to obtain needed nutrition. Although we cannot say with any certainty that other carnivorous mammals possessed similar adaptations to an herbivorous diet in the past, the anatomy and physiology of the panda does illustrate the possibility that modern carnivores could have survived as herbivores.

10.3.4. Mechanism: Ebola

The general phenomenon of emerging diseases offers an excellent example of the deterioration of creation as a result of the Fall. To an evolutionist, an emerging disease represents another chapter in the ongoing struggle between pathogens and hosts. As baraminologists, we hypothesize that newly emerged diseases are recent deteriorations of otherwise harmless creations. If we accept that creation is degenerating as a result of the Fall, we should expect to see continuous evidence of that degeneration throughout history. Specifically, we should find that the virulence of emerging pathogens will vary widely among closely-related species, perhaps even forming a continuum from harmless symbiont to fully virulent pathogen.

Though not technically an organism, the ebola virus provides an excellent example of an emergent disease that varies widely in pathogenicity. The various strains of ebola, together with the Marburg virus, belong to the virus family Filoviridae. Marburg appeared first in 1967, followed by the first cases of ebola in 1976. Ebola infection manifests itself by a severe hemorrhagic fever after a week-long incubation. The fever often exceeds 101°F, and bleeding occurs from all mucous membranes. These symptoms can persist for three weeks before the victim succumbs. Though virulence varies among the different strains of ebola, the overall fatality rate exceeds 70 percent.[8]

Despite a quarter century of research, scientists still do not fully understand the biochemistry and natural history of the filoviruses. In particular, no one has been able to identify the reservoir of these viruses.[9] Based on genomic evidence, we can confidently say that Marburg and the various ebola strains do share significant continuity.[10] Most importantly, as we mentioned above, the virulence of filoviruses varies from strain to strain. Ebola Zaire has the highest fatality rate, claiming 87 percent of its victims in the first outbreak in 1976. In contrast, the worst outbreak of Ebola Sudan killed only 64 percent of patients, and Marburg has a fatality rate of just 24 percent.[11] Most surprisingly, Ebola Reston, the strain that inspired the book *The Hot Zone,* produced no symptoms in humans.[12] Given a gradually degenerating creation, we would expect precisely the kind of variation in virulence among emerging pathogens that we observe in the filoviruses.

10.3.5. Rate: Dinosaur carnivory

As we noted above, creationists often explain modern predation by using an evolutionary mechanism, in which the future predator gradually changes from an herbivore to a scavenger and ultimately a predator. According to this scenario, active hunting for prey probably did not come about until after the Flood. By the time animals turned to active predation, many carnivores, such as the dinosaurs, had undoubtedly gone extinct. This hypothesis appears consistent with our understanding of a degenerating creation, but it has been neither tested nor supported by historical evidence. The hypothesis would be invalidated by direct evidence of active predation in pre-Flood organisms. If we find evidence of pre-Flood predation or associated aggressive behavior, then we would have evidence to support the rapid emergence of carnivory. Since dinosaur fossils were deposited during the Flood, the fossilized dinosaurs initially lived prior to the Flood. Thus, examination of dinosaur bones and other remains can provide evidence of conditions that existed before the Flood.

There are numerous ways of studying the diet and feeding habits of dinosaurs based only on fossil evidence. For example, careful studies of the potential bite force of *Tyrannosaurus rex* and *Allosaurus fragilis* have revealed important details about the hunting behaviors that these two carnivorous dinosaurs were capable of performing. Because of the weak bite of *Allosaurus,* paleontologists believe that it must have made quick bites in order to induce bleeding in its victim, leading to the animal's death. In contrast, a single bite from *Tyrannosaurus* could have killed immediately because the *T. rex* skull allowed for a much stronger force.[13] Similarly, an examination of a *T. rex* coprolite (fossilized feces) revealed a very high percentage of tiny bone fragments, indicating the large amount of meat eaten by this giant carnivore.[14] These observations alone do not provide unambiguous evidence of pre-Flood predation since they could result from scavenging as well.

Direct evidence of pre-Flood predation comes from paleopathology. For example, although tooth marks in bone could originate after the organism was dead, tooth marks that show evidence of healing must have occurred when the animal was still alive, indicating that it survived an attack by another animal. The largest *Tyrannosaurus rex* skeleton discovered to date, the Field Museum's "Sue," bears evidence of vicious aggression, presumably by other tyrannosaurs. Ribs on both sides of Sue's body were broken and healed in life, but on the left side, one rib only partially healed possibly because of a tooth fragment left in the wounded bone. Healed holes on her lower jaw testify to bites from other tyrannosaurs, possibly during mating season or territorial disputes.[15] While these evidences do not unequivocally testify to a predator, they strongly imply the kind of aggressive behavior frequently associated with carnivores that hunt.

10.3.6. Rate: Diseases

In addition to carnivory, many creationists also believe that diseases emerge gradually. Previously, we argued that ebola and emerging diseases support the model of a deteriorating creation. What we need to know now is whether disease emergence has always been slow and gradual or whether the rate of disease emergence was different in the past. Because the age of a particular disease raises questions about its survival through the Flood, we need good studies and models of disease emergence to understand how pathogens and parasites weathered the worst catastrophe in history.

A survey of Genesis reveals no biblical evidence of diseases prior to the Flood. The earliest diseases mentioned in the Bible are Job's boils (Job 2:7), which he describes with some detail throughout the book (Job 7:5; 13:28; 30:17–18, 30). The details of Job's illness leave little doubt that he was infected by some kind of pathogenic bacteria, but we cannot say for certain what particular kind of bacteria he had. At about the same time "plagues" were inflicted on Pharaoh after taking Abram's wife (Gen. 12:17), but the terminology in this passage is not used exclusively of disease. Without a more detailed description of Pharaoh's afflictions, we cannot conclude that this was related to disease.

Infectious diseases are difficult to diagnose from skeletal remains, but evidence of dinosaur infections exist for several species.[16] Once again, since dinosaur fossils preserve remains of pre-Flood organisms, we know that bacterial infections must have existed before the Flood. Whether these infections were caused by the same bacteria that cause modern diseases cannot be determined unequivocally. Thus, we might conclude that bacterial diseases emerged rapidly after the Fall because of their occurrence before the Flood, or we could hypothesize that new diseases have been emerging throughout history. Either of these interpretations would be consistent with the present evidence.

10.4. Imperfection and Anthrax

Baraminologists studying imperfection face the largest challenge in teasing apart the different layers and levels of imperfection from what might seem like a simple problem. To demonstrate the layering of imperfection, we here present an extended example of an apparently severe biological imperfection: anthrax.[17] Anthrax provides an excellent model of the multilayered complexity of biological imperfection. Anthrax disease results from infections by the bacterium *Bacillus anthracis*. As long as the bacteria do not spread to the bloodstream, the infection is relatively benign, as in gastrointestinal or skin anthrax. When a person inhales *B. anthracis* spores, however, the spores quickly germinate in the lymph nodes of the chest and invade the blood. In the blood, the actively-growing *B. anthracis* bacteria begin producing a dangerous toxin called lethal toxin that kills with amazing efficiency. Within three to five days of infection through inhalation, most patients die. The onset of death comes so quickly that effective diagnosis and treatment is nearly impossible. By the time the physician realizes the cause of the initial flu-like symptoms, the patient is frequently beyond successful treatment.

The genes encoding the components of the *B. anthracis* lethal toxin are found on a plasmid, a circular piece of DNA that replicates independently of the main bacterial chromosome. A second anthrax plasmid gives the spores a tough outer coating that resists attack by the host immune system. If a single *B. anthracis* bacterium lost either the toxin plasmid or the coat plasmid, that bacterium would be incapable of causing anthrax disease. This illuminates the first level of complexity in explaining the origin of anthrax disease: We can separate the origin of the bacterium from the origin of the pathogenic plasmids.

An important principle of our analysis of imperfection is degeneration. If the origin of the anthrax bacterium came about by degeneration, we ought to be able to demonstrate that the progenitors of modern *B. anthracis* were harmless. *B. anthracis* belongs to a group of bacteria that normally live freely in the soil. Specifically, *B. anthracis* belongs to a small group of six *Bacillus* species that are difficult to distinguish. Several reports have shown that *B. anthracis* cannot be distinguished from other species when it loses its plasmids. This similarity surely counts as significant, holistic similarity, placing anthrax into the same baramin as harmless soil bacteria. For anthrax then, we find ample evidence supporting the origin of pathogenicity in the anthrax bacterium from an otherwise harmless organism by the acquisition of two plasmids.

At this stage, we have explained one layer of the complex origin of anthrax disease. *B. anthracis* became the deadly anthrax bacterium by acquiring a plasmid that produces toxins and a plasmid that protects the bacterium in the human bloodstream. Now we turn to the next layer of complexity–where did the harmful plasmids come from? We could answer this question by trying to identify signs of degeneration in the plasmids and therefore propose that these plasmids arose accidentally from originally beneficial sets of genes. The plasmid carrying the genes that confer resistance to the immune system need not be examined in such a way. By itself, this plasmid only allows bacteria to survive in the human body. If the bacteria carry no toxin genes, then they would not cause disease. In fact, it is possible to imagine a symbiotic relationship between humans and bacteria, in which the bacteria protected by an outer coat provide a benefit to the human host. Such bacterial symbioses are commonplace; thus, the existence of a plasmid that provides protection from a host immune system fits easily into a benevolent creation.

In contrast to the protective plasmid, the toxin plasmid poses a much more serious challenge to the creation model. The complete DNA sequence of the anthrax toxin plasmid provides an excellent resource to study the origin of the toxin genes. One of the first things we find in the plasmid sequence is evidence of deterioration of the plasmid itself. In contrast to the long genes of the *B. anthracis* chromosome, the plasmid genes are much shorter, probably because of mutations that truncate them, further evidence of a gradual breakdown of order. Most importantly, the genes that encode the anthrax toxins are found in a region of the plasmid called a *pathogenicity island.*

In bacterial genomes, genes associated with disease and toxicity often occur in physical proximity on the chromosome, in a region bounded on either end by repetitive or insertion sequences. These regions often show a great deal of sequence variation even among closely related isolates of the same bacteria. Some strains even lack the region altogether. Because of this variety in sequence and because of the importance of these regions to disease, scientists call them pathogenicity islands. Because of their propensity to be absent from some closely related strains or species of bacteria, pathogenecity islands are often classified among the mobile genetic elements, pieces of DNA that can move around independently in organismal genomes.

For our discussion of anthrax, the existence of a pathogenicity island on the toxin plasmid in *B. anthracis* adds an exciting new layer of complexity to the problem of anthrax imperfection. We noted above that the anthrax pathogen could be explained as a harmless soil bacterium that acquired a toxin plasmid, and now we find that the toxin plasmid could be explained as a harmless plasmid that acquired a pathogenicity island. This amazing pattern highlights the recurring theme of modularity that dominates both design and imperfection. The bacterium *B. anthracis* became pathogenic by acquiring the modular plasmid, and the plasmid became toxic by acquiring the modular pathogenicity island. Of course, the next question we ought to ask should be readily apparent: If the plasmid became imperfect by acquiring a pathogenicity island, where did the pathogenicity island come from?

To identify the origin of the pathogenicity island, we must now examine the possible origins of the underlying genes. Three toxin genes occur in the pathogenicity island, one encoding a protective antigen (PA), one encoding edema factor (EF), and one encoding lethal factor (LF). After synthesis of these proteins, the bacteria secrete them into the bloodstream. The PA and EF proteins together form edema toxin, the function of which is unclear. PA and LF together comprise the lethal toxin. When injected directly into the bloodstream, lethal toxin (PA and LF) alone can kill mice with the same deadly efficiency as fully virulent anthrax spores.

The chemical structure of PA is highly unusual in that seven separate PA proteins assemble together to form the fully functional PA protein complex, which binds to the surface of human cells. While six-unit protein complexes (hexamers) are common in living organisms, seven-unit protein complexes (heptamers) are rare. Binding of one unit of either LF or EF to the PA heptamer forms the mature toxin complex. The toxin complex induces a structural change in PA, which in turn stimulates the cell to engulf the toxin complex by the process of phagocytosis. At this point, PA facilitates the injection of EF or LF directly into the cytoplasm of the cell, where these proteins then interfere with normal cell signaling.

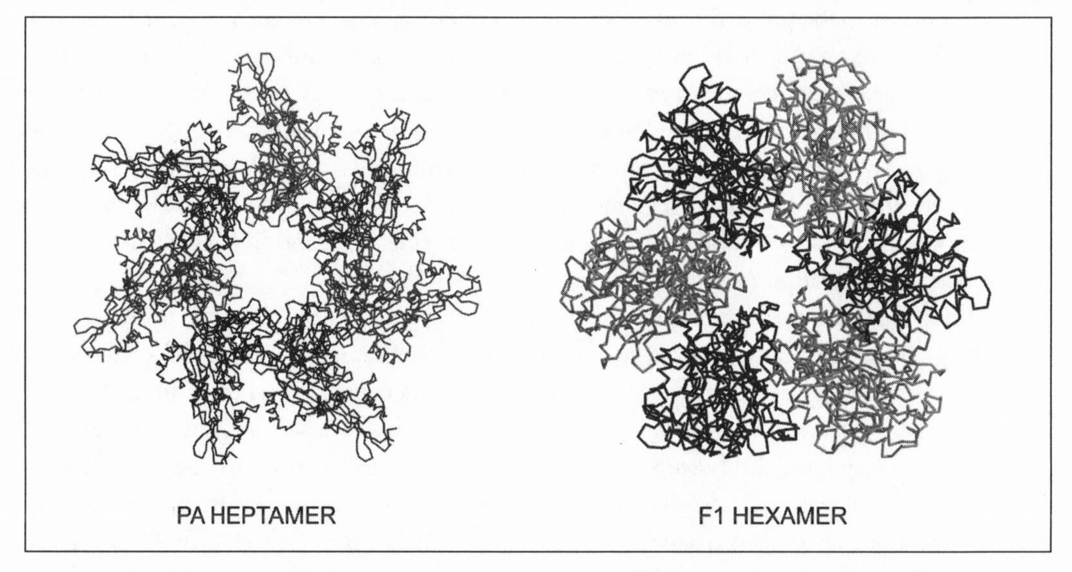

Figure 10.3. *The molecular structure of the anthrax protective anthrax (PA, left) is composed of a seven-member ring of proteins. Six-member rings like the F_1 ATP synthase on the right are much more common. The complex structure of PA implicates design in the origin of anthrax.*

Considering the complexity of the process of secretion and activation of the toxin complex, proposing a degradative origin for these genes and proteins would strain credibility. The genes must have been designed, but for what purpose remains a mystery. Because much research remains before we fully understand the actions of LF and EF within the cell, we should not yet draw any firm conclusions about the purpose of LF or EF in an unfallen world. Curiously, the proteins involved in anthrax toxins bear some resemblance to toxin proteins from other bacteria, including bacteria that cause tetanus, botulism, and food poisoning. These similarities may hold important clues to understanding the original purpose of bacterial toxin genes and will certainly allow the creationist explanation of anthrax toxins to be applied to other bacterial toxins.

10.5. Chapter Summary

Man has long struggled with reconciling a benevolent, omnipotent God with the existence of evil. We often hear the question "If God is a perfect designer, why is His design not perfect?" Examples of imperfection include vestigial organs, junk DNA, and disease.

Imperfection begins with the groaning of creation that resulted from the Fall of man. As we see in Genesis 3, the Bible clearly teaches that something is wrong with God's creation. It no longer occupies that "very good" state in which God created it. Other claimed imperfections, such as the vertebrate retina, may not be imperfect at all. Instead, our perception of God and design is simply inadequate.

We find some imperfections that relate primarily to the behavior or ecology of organisms, in the broadest sense. We also find a more challenging form of imperfection in structures or substances that seem to be designed to harm other organisms. Examples include the complex multiprotein toxins produced by some bacteria, the poisonous substances produced by some plants, and the amazing venom-injection system of snakes. There is a third category of imperfections that arise because of degeneration from the originally perfect creation. Examples of degeneration include mutations that lead to such genetic diseases as cystic fibrosis.

Review Questions

1. Explain why biological imperfection is preferred to natural evil.
2. How did Stephen Jay Gould misinterpret the imperfection of orchids and pandas?
3. What is a pathogenicity island? Illustrate their importance to imperfection by using anthrax as an example.
4. What is wrong with creation?
5. Name the three cursed individuals from Genesis 3. What are the biological implications of the Curse for each individual?
6. What can we infer about the Curse from Isaiah 11?
7. Does all imperfection come from the Curse?
8. How does baraminology help us understand imperfection?
9. Why does the rate of the appearance of imperfect impact our understanding of historical biology?
10. Explain Dawkins's argument against design by using the vertebrate retina.
11. Illustrate a continuum of diet by using the members of the bear family Ursidae.
12. How is the giant panda a carnivore?
13. Give evidence of pre-Flood dinosaur carnivory.
14. What biblical or baraminological data can help us to determine the difference between true imperfections and curious facets of design?
15. List and describe three examples of imperfections attributed by creationists to the Fall and the Curse.

For Further Discussion

1. Develop criteria that could be used to identify imperfection.
2. Explain how imperfections can be used to support the creation model and the evolution model.

3. Besides the examples given in this chapter, think of another example of imperfection as a result of degradation.
4. Has the appearance of imperfection increased, decreased, or remained the same since the Fall?
5. Give other potential mechanisms of imperfection besides the ones discussed in this chapter.
6. Explain the implications of dinosaur carnivory for pre-Flood humans.
7. Assuming a degradation model of imperfection, speculate on the possible purposes that anthrax toxins originally had.
8. If it was possible to completely eradicate bacterial toxins, would it be theologically or biblically justifiable to do so?
9. If it was possible to completely eradicate pathogenic bacteria, would it be theologically or biblically justifiable to do so?
10. If the Fall had never happened, would humans still have vestigial organs?

CHAPTER 11

Baraminology and Diversification

11.1. Introduction

Because of Linnaeus's lasting influence, the question of species fixity and transformism remains contentious among some creationists to this day. Many creationists seem to accept species fixity with no speciation at all, while other creationists uncritically accept neodarwinian mechanisms of speciation, so long as the changes stay within the perceived "boundaries" of the baramin. Still others apply the imperfection principle of degeneration to the question of speciation, insisting that all evolution necessarily results in the degradation of the "perfect," created population.

Beyond the banal question of whether speciation could occur, modern phylogenetic studies make certain assumptions that may not be true in the biblical framework. For example, most popular phylogenetic methods disallow reticulate evolution, in which branches of the evolutionary tree re-merge after divergence. Reticulate evolution can occur through horizontal gene transfer or hybridization. Since hybridization occurs so frequently, as we saw in chapter 7, we should not underestimate its importance in species transformation. The various cladistics methods also make questionable assumptions regarding the mechanism of evolution. These methods assume that evolution is **parsimonious**, that is, complex characters evolve only once and are passed on from parent to child. With a high frequency of hybridization, traits could easily introgress, or pass from one species to another without proceeding through a common ancestor. Consequently, marrying baraminology to modern neodarwinism is certainly unwarranted.

Mindful of the problem of inferring species ancestry, we prefer a definition of baramin (and related terms) that is purely descriptive, not relying on phylogenetic inference of ancestry. In the preceding chapters, we spoke of speciation in general terms of "change," but in this chapter, we must return to the issue of species transformation to try to develop an understanding of the subject consistent with our baramin concept and biblical history.

Once again, rather than adopting evolutionary concepts to baraminology, we strongly recommend the use of uniquely creationist terminology to describe species transformation. Evolutionists have a number of terms for fine-scale evolutionary changes. **Microevolution** refers to changes that occur within a species, sometimes generating subspecies, varieties, or races. **Speciation** is the process by which new species arise. Evolutionists usually use **macroevolution** or megaevolution to refer to evolutionary changes that result in new genera, families, orders, and even larger groups. If baramins encompass more than one genus (which they often do), then organismal transformations within a baramin could legitimately be described by all three evolutionary terms, depending on the diversity of the resulting species. We use the term intrabaraminic **diversification** to describe the generic process by which members of a baramin bring forth new species or varieties, without regard to the magnitude of the change.[1]

As a scientific term, diversification is free from the theoretical and philosophical baggage of the evolutionary terms mentioned above. Consequently, we can use *diversification* while we further develop and refine our understanding of baraminological changes. Over time, we should come to appreciate better the rate(s), pattern(s), and mechanism(s) of diversification. This understanding will help us to develop a better model of how God designed baramins to be capable of generating diversity. In the following sections, we will review our present understanding of these three important considerations regarding diversification (rate, pattern, and mechanism). We will conclude the chapter with a discussion of the relationship of diversification to design.

11.2. Diversification Rate

Wise devised an analogy of an orchard to illustrate his baramin.[2] According to the analogy, each tree of the orchard represents a baramin, with the branching of the tree representing diversification. The orchard contrasts with the single tree of evolution, which contains all diversity, signifying that all species descend from a single ancestor, and the "lawn" of Linnaeus, in which no branching occurs at all, signifying that species do not change. In terms of the pattern and rate of diversification, the selection of a "tree" analogy gives the impression that diversification does not significantly differ from evolution. Just like the single evolutionary tree of life grows gradually and branches continuously, so too the baraminological trees grow gradually with continuous branching.

More recent baraminological research has helped to clarify the pattern and rate of diversification in vertebrate baramins. The picture emerging is one of rapid initial diversification after the Flood, followed by a prolonged period of relative stasis.[3] We infer the period of rapid diversification from the biblical record and archaeological studies that supplement our understanding of baramins. We readily find records and remains of modern species very early in post-Flood history. Because these species belong to baramins with many other species, we know that they must have diversified from their ark-born ancestors very quickly after (or even during) the Flood.

We have already mentioned the gifts given by Pharaoh to Abram as recorded in Genesis 12:16. These gifts included sheep, oxen, camels, and donkeys. Two of these species

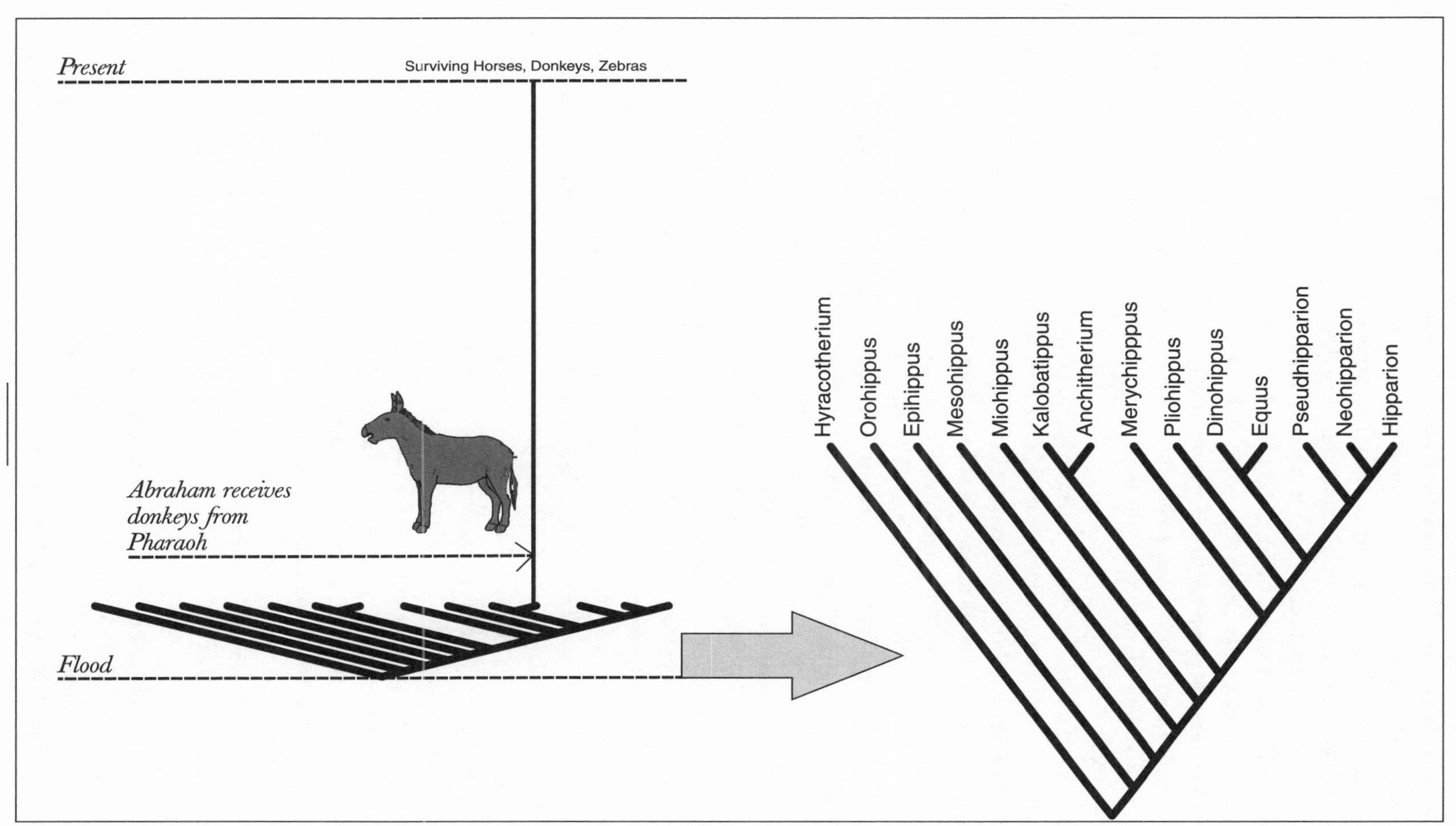

Figure 11.1. *Because of the Bible, we know that post-Flood diversification must have been very rapid. Since Abram received donkeys from Pharaoh only 370 years after the Flood, all of horse diversification (shown on the right) must have taken place prior to that event.*

have been evaluated in a baraminological context: camels and donkeys. The camel here is most likely the dromedary that is so common in the modern Middle East.[4] As we pointed out in chapter 7, a dromedary and guanico have been successfully crossed, producing a living hybrid offspring. This hybridization unites two distinct groups in the family Camelidae, implying that all camelids belong to the same baramin. The fossil record of camelids consists of 200 species, all of which occur in post-Flood sediments. Since modern dromedaries are one of the most recent species in the history of the camelid baramin, these 200 other camelid species must have arisen prior to the origin of the dromedary. Since we know that Pharaoh and Abram met around 370 years after the Flood (by the genealogical chronology in Gen. 11), those 200 camelid species must have arisen, left fossils, and probably went extinct in less than 370 years.

A more forceful case for rapid diversification can be made for the horses of family Equidae. A recent baraminology study has covered nineteen different fossil equid species, including the ancestral *Hyracotherium,* as we have described in previous chapters.[5] The results of this study indicate that all nineteen fossil equid taxa belong to a single monobaramin. This evidence directly places fossil taxa in the same baramin as modern taxa, and by similarity to other fossil species, we can infer at least 150 fossil horse species in the same baramin. Again, because most of these fossil equid species precede our modern *Equus* (including donkeys), these species must have originated in just 370 years immediately after the Flood.

Job confirms the early occurrence of many modern species. Job refers to horses (39:18–19), and he was a keeper of donkeys, sheep, oxen, and camels (1:3). This information directly corroborates the existence of these species within centuries after the Flood. Additionally, Job also refers to lions (Job 4:10–11; 10:16; 28:8; 38:39), dogs (30:1), and three different birds of prey: eagles (9:26; 39:27), hawks (39:26), and vultures (28:7). As with the camels and horses, the lions and dogs also represent very diverse baramins that must have largely finished their diversification very quickly after the Flood.[6] In addition to these mammalian baramins, the direct mention of three different birds of prey indicates that rapid diversification also took place in three different monobaramins of birds.[7]

As a supplement to these biblical studies, we can also examine archaeological remains, particularly artwork that depict modern species. Early human paintings and sculptures frequently portray animals that can be identified with modern species. Although some cave paintings depict extinct creatures such as woolly mammoths, no representations of the earliest pre-Flood horse, camel, cat, or dog species are known to exist. We may attribute this lack to two important reasons. First, these early species emerged and died out so rapidly that they made little contact with people. Second, most people remained in the region around Babel during most of the period of diversification, thus avoiding contact with exotic animals. By the time people dispersed from Babel and encountered other animals, diversification had produced a wide variety of species and the Decimation (discussed in chap. 3) had already wiped them out.

Taken together, all of these observations point to the conclusion that the Flood was immediately followed by a period of explosive diversification. These same observations also indicate that the period of diversification was followed by an extended period of stasis. Most species mentioned in the earliest biblical passages and depicted in early human artwork can be identified with living species, which must have remained unchanged from the time of Abraham, some four thousand years ago. Thus, Linnaeus and his contemporaries were partially correct. Species do appear stable over time. Linnaeus erred by assuming that what he observed in the present (species stasis) must necessarily hold for all time (species fixity).

11.3. Diversification Pattern

By diversification pattern, we mean a reconstruction of the historical pathway taken by a baramin through diversification. The widely accepted pattern of evolution is a simple tree structure, in which extant species appear as twigs or leaves and ancestral species appear as limbs, trunks, and roots. As we have already noted, creationists are fond of "accepting" evolution as long as it stays within the limits of the potentiality region. This attitude has led creationist biologists to assume that evolutionary studies and phylogenetic inferences are acceptable when members of a single baramin are being analyzed.

The primary assumption that undergirds all evolutionary studies is common ancestry. Thus, if we have a group of organisms that truly share a common ancestor, they fulfill that primary assumption. In Section 11.1, we briefly introduced other methodological assumptions that are not necessarily met within a baramin, and we suggested that baraminologists avoid equating diversification with neodarwinian microevolution and speciation. In this section, we will further develop the importance of developing a uniquely baraminological perspective of diversification that does not rely on adapting evolutionary phylogenetics.

One foundational assumption in all evolutionary studies is the idea that degree of similarity reflects degree of kinship. Accordingly, organisms that are very similar are assumed to be closely related, and organisms that are dissimilar are assumed to be more distantly related. Various features of evolution make this assumption inescapable. First, evolution proceeds very gradually, building up small differences over time. Even in the punctuated equilibria model with its rapid speciation changes, the new species are never radically different from their parents. As a result, distantly related species tend to show more differences than closely related species. As we just argued in the previous section, diversification did not proceed in a slow and gradual fashion. Instead, diversification occurred in one short burst, followed by an extended period of stasis. Consequently, we should exercise caution when accepting the phylogenetic link between similarity and kinship.

The second reason that evolutionists believe that similarity reflects kinship is the rarity of convergent evolution. When two different species acquire a similar feature independently, this is called *convergent evolution*. Under the evolutionary model, new traits come about by chance mutations or alterations in preexisting material (body plans, genes, etc.). Because the origin of a complicated new trait is highly unlikely, a second origin of that same trait is even more unlikely. Consequently, convergence must be rare within the evolutionary model.

As we argued in chapter 9, baraminology gives us a new way of understanding how God mediates His design through diversification. The most important implication of mediated design is the rejection of the random origin of new traits within baramins. If diversification and adaptation were part of God's original plan, then they do not happen by chance. As a result, convergence may not be rare at all, and similarity might reflect only general kinship trends, not specific relationships. For example, similarities that warrant classification into different tribes or families may indeed indicate a distant relationship, but the similarities among species of the same genera may be poor guides to kinship.

On a more technical level, phylogenetic methods assume two dubious ideas: bifurcating trees and parsimonious evolution. All computational methods that generate evolutionary trees assume that the evolutionary pattern is best described by a bifurcating tree, in which a species can split into only two species. In the creation model, with so many new species arising in such a short time, baraminologists should not limit themselves to such a restrictive structure. It may be possible that several species arose from a single population (resulting in a "multifurcation" or polytomy). It may also be possible that species bifurcated so quickly that the actual pattern of divergence is unresolvable.

The principle of parsimony directs the tree calculations in cladistics and related methods. The principle states that the phylogenetic hypothesis (tree) that requires the fewest changes (mutations, origins of novel structures, etc.) is preferred over trees that require more changes. We can see similarities between parsimony and Occam's razor ("the simplest theory is most likely correct"), but the principle also has roots in the idea that complex traits are unlikely to arise more than once. As a result, different species that possess a particular trait most likely acquired that trait from a common ancestor in which the trait originated.

Under baraminology with mediated design stimulating the expression of pre-designed characteristics, there is no compelling reason to believe that diversification is parsimonious. Since the characteristics already exist, the probability for any two species within a baramin to express the same characteristic would be very much higher than if they had to evolve those characteristics from scratch. Thus, we should not expect diversification to be particularly parsimonious.

If diversification does not follow the pattern of evolution, what pattern does it follow? Instead of a slow-growing, neatly-branching orchard of trees, we can think of baramins and their diversity as a fireworks display. In a professional fireworks show, shells are shot from a mortar and explode at a predetermined time. From the shell, small packets of combustibles called "stars" emerge in a burst, sometimes leaving an illuminated trail behind them. We can view a baramin or holobaramin as a single lineage following a particular historical path until the Flood. This initial pathway of the baramin is represented by the unexploded shell launched from the mortar. After the Flood, the baramin rapidly diversifies into many different species, represented by the burst of the shell into its many different stars. The species then either go extinct or remain stable until the present time, in the same way that the stars that burst from the shell follow a single trajectory.

Our view of the rate and pattern of diversification also has a biological parallel: animal development. For this analogy, we will view an adult animal as a collection of diverse cell types, each of which represents a different species of a baramin. Each of these cell types arose from a single progenitor, the zygote, comparable to the common ancestor of many extant baramins. The zygote divides, grows, and becomes all the different cell types in an adult body through the process of development. Compared to the normal lifespan of an adult organism, the development period is often very short, as is the process of diversification.

Figure 11.2. Considering all attributes of diversification discussed in the text, we can liken the process to a fireworks display. Like stars shooting out from a shell, species radiate rapidly from the baraminic members that survived the Flood.

Once the adult form is reached, the cells normally do not differentiate into other cell types, just as species stop diversifying after a fixed period of time.

This view of baraminic diversification has a number of important implications. The first major implication concerns the branching pattern. It is unlikely that anyone would be able to reconstruct the packing of the stars in the original shell simply by examining the trajectory of the stars. The only pattern that can be described is the overall pattern produced by the explosion of the shell. In development, it is possible to trace the differentiation and growth of individual cells, but only because we can repeatedly observe development in controlled situations. Through these types of experiments, we have found that the resulting development pattern is not a simple tree. Adult tissues come from different progenitor cells, leading to an elaborate network pattern.[8] As we evaluate baraminic diversification, we must keep in mind that detailed phylogenetic relationships between species may be unresolvable.

The starburst pattern of diversification also implies that studies of speciation and reproductive isolation in the modern world probably have little to do with the initial species formation during diversification. In a fireworks display, the minor variations in star pathways are determined by gravity and wind direction, not by the shell explosion that produced them in the first place. Within development, differentiation and development are very different processes than the processes that maintain the adult body. Similarly, the minor variations we observe today in baramins have very little effect in producing the kind of diversity that we observe in the whole of the baramin.

We might observe limited size or color differentiation between two distinct populations of a single species, but we rarely observe differences like that between the lion and tiger or between the dog and fox. Since baraminic diversification is responsible for all modern felid and canid species, this kind of diversity is exactly what must be generated. Consequently, the mechanisms generating diversity in modern species are not the mechanisms that produced the original species after the Flood. In terms of our fireworks analogy, fluctuations in the trajectory of stars cannot be attributed to the explosion of the shell, nor can the explosion of the shell be attributed to the trajectory of the stars. This brings us to the final major feature of diversification: the mechanism.

11.4. Diversification Mechanism

Partly due to Marsh's influence, creationists came to accept conventional speciation mechanisms to explain intrabaraminic diversification. As we have just argued, the rate and pattern of diversification precludes adoption of these mechanisms. Modern variation and speciation produce diversity orders of magnitude less than the diversity following the Flood. The mechanism of diversification must be qualitatively different from the mechanisms of speciation today. To discover possible diversification mechanisms, we can begin by inferring mechanistic properties from the rate and pattern of diversification, then seek mechanisms that display these attributes.

11.4.1. Attributes of the Diversification Mechanism

11.4.1.1. Design. We have already alluded to the first property of the diversification mechanism: design.[9] Under baraminology, adaptation and diversification are not natural responses to environmental changes. Diversification is designed by God to allow His revelation to persist in a changing world. Thus, diversification can be viewed as the expression of attributes that God designed for each baramin. These unexpressed attributes can be built into creation in a variety of ways, which we will discuss in more detail below.

As a preliminary example, consider recessive traits, such as dwarfism in peas. Occasionally, a dwarf pea plant will be born to two pea plants of normal height. The reason for this seeming incongruity is that the attributes of some alleles are simply not expressed when in the presence of alleles that dominate the phenotype. Alleles that "defer" to the attributes of others are called recessive. Only when two of the same kind of recessive alleles appear in the same individual is the recessive trait expressed. In the case of pea height, the normal height allele is dominant over the recessive dwarf allele. Thus, any population of peas of normal height could be carrying the allele necessary for dwarfism. Although unexpressed traits can be encoded in many other important ways, recessive alleles effectively illustrate the potential for complex traits to arise by mechanisms that easily entail design.

11.4.1.2. Specificity. The second property that the diversification mechanism must exhibit is specificity. Under all neodarwinian speciation mechanisms, novel features can only arise by chance. Because complicated features have a low probability of originating by chance, neodarwinian speciation necessarily requires long periods of time to produce the kind of variation we observe in modern baramins. Since we know that diversification accomplished the same task in a very short amount of time, the mechanism must not be based on chance. Not only is the probability of generating novel characteristics much higher in diversification because the traits are designed, but the mechanism that activates those characteristics must work with specificity.

For example, imagine that one gene in a hypothetical plant baramin allows the individual plants to grow in saltwater marshes. Now imagine that in most members of this hypothetical baramin that gene was inactive because of a single nucleotide change. If this were the case, the activation of that gene (and the expression of the saltwater-growth trait) would require a precise point mutation in the gene, leaving us with little speed advantage over neodarwinian speciation. The probability of the correct mutation at that particular nucleotide is astronomically low, so we would have to wait a long time before the saltwater-growth characteristic was activated, if it became active at all.

Alternatively, if God created that nucleotide position to be highly mutable (a "hotspot"), the probability of expressing the saltwater-growth trait would be much higher, and the trait would be expressed much more quickly. The difference is one of specificity. If the critical nucleotide is just like every other nucleotide, expression of the saltwater-growth trait proceeds slowly, but if the nucleotide is one specifically designed to change, expression of the trait can occur rapidly.

11.4.1.3. Stability. Since we currently observe species stability, we can also infer that the changes wrought by diversification must themselves be stable. Although scientists have observed living organisms for centuries, no one has observed the kind of rapid diversification that took place after the Flood. Instead, modern species demonstrate a remarkable reproductive fidelity, particularly among the vertebrates. We do not find elephants giving birth to mammoths or lions giving birth to tigers, yet changes of that magnitude may well have accompanied births during diversification. Whatever mechanism caused diversification in the first place must have rendered permanent changes in the nascent species. As a result, elephants give birth to elephants, and lions give birth to lions in our modern world.

Because of the permanence of diversification, modern efforts to breed new varieties of plants and animals actually serve as poor examples of diversification. For example, the varieties of the plant species *Brassica oleracea* include cabbage, kale, broccoli, and cauliflower. Although each of these vegetables appear very different, they were all bred in recent history from a single variety of *B. oleracea*. Despite their apparent differences, they remain the same species and if given the opportunity would interbreed to become a single variety once again. In contrast, the species that arose by diversification do not interbreed but remain separate and stable species, especially among the animals. Although modern breeds and varieties can effectively illustrate the phenotypic potential inherent in a species, the instability of breeds and varieties provides poor evidence of diversification.

11.4.1.4. Modern inactivity. Finally, the rapid cessation of diversification after the Flood implies that the diversification mechanism must be inactive in the modern world. As we explained in the previous section, we do not observe diversification in our present world.

Members of most species reproduce other members of the same species. The phenotypic diversity in reproduction so common during diversification simply does not occur today. Consequently, we conclude that whatever mechanism that generated diversification during the post-Flood recovery period is no longer operating to the same degree it did then.

The modern cessation of diversification raises an important question: Did diversification cease forever or is it merely dormant? A permanent cessation of diversification means that it could not happen again, regardless of the biological conditions. A dormant diversification mechanism means that diversification could happen again if the correct conditions triggered its activity. Although a permanently inactive diversification explains the biological and historical data as well as a dormant one, we can make a better theological case for dormancy.

To review from chapter 2, we attribute intrabaraminic diversification to God's desire to see His revelation persist. Thus, He designed organisms with the ability to adapt during radical changes in environmental conditions, such as would accompany a global deluge. If He truly wanted His baramins to persist, He would certainly create a mechanism that would also persist. Permanent cessation of diversification implies that the mechanism of diversification has not actively persisted, endangering God's revelation in the baramins. Diversification dormancy implies a higher-order design, yielding a mechanism capable of persisting.

11.4.2. Mechanistic Explanations of Diversification

Having established the characteristics of the diversification mechanism (designed, specific, permanent, and currently inactive), we may now turn to various phenomena of the biological world to determine if any of them fit the description. Creationists have proposed several mechanisms to explain intrabaraminic diversification, including neodarwinian speciation, a Mendelian mechanism called **heterozygous fractionation**, and a mechanism derived from genomic studies called **genomic modularity**. As we will see, none of these mechanisms fully explain all of the necessary attributes. Understanding diversification remains an active and exciting area of creationist research.

11.4.2.1. Neodarwinian speciation. Perhaps out of convenience, acceptance of neodarwinian microevolution and speciation remains a common feature of creationist writing about the post-Flood diversification. Having established the characteristics of diversification and its mechanism, we can see now that diversification is far too fast, specific, and permanent to be generated by neodarwinism. This is not to say that neodarwinian mechanisms do not operate; they certainly do.

As we explained in our fireworks example, a shell represents the baramin, and the stars that emerge upon ignition of the shell represent species. Variations in the trajectory of the stars occur because of the combination of gravity and wind. Sometimes stars can even fragment as they burn, yielding two stars where there was once only one. This type of variation represents neodarwinian variation. The variation is minor compared to the initial diversification process, and the changes wrought are generally small compared to the diversity of the entire baramin. While we cannot discount neodarwinism as a real biological phenomenon in our modern world, we cannot accept neodarwinism as the causative agent of diversification.

11.4.2.2. Heterozygous fractionation.[10] Despite Marsh's endorsement of mostly neodarwinian mechanisms of intrabaraminic diversification, most modern creationists advocate a mechanism that differs considerably from Marsh's. According to the heterozygous fractionation model, baramins began with a large gene pool with many different alleles, all created by God. In other words, the individual organisms created by God were heterozygous at most of their gene loci. As time progressed, that original gene pool became fragmented or fractionated into smaller pieces. The process of fragmentation produced the variety of species we have today.

Whereas heterozygous fractionation could be considered a design feature, it fails to explain the specificity, speed, and current dormancy of diversification. If we consider just vertebrate baramins, we quickly realize that the gene pool could not have been very large at the Flood. Assuming two of each unclean baramin survived the Flood aboard the ark, only four alleles per locus at most could have been preserved. Since the Flood occurred nearly fifteen hundred years after Creation, the number of alleles per locus is probably much lower than four, since the fractionation of the created gene pool would already have commenced. Since we know that modern members of baramins have many more alleles per locus than just four, heterozygous fractionation actually requires an additional mechanism to generate allelic diversity upon which it can act. Having only a few alleles per locus after the Flood would not have produced the rapid diversification that must have occurred.

Additionally, heterozygous fractionation suffers from an irreversibility that renders it a temporary mechanism at best. Since allelic diversity is attributed to Creation, once that gene pool has fractionated, it cannot be reformed and refractionated without unmediated divine activity (unless an additional mechanism exists to generate allelic diversity). Thus, the mechanism of heterozygous fractionation cannot persist.

11.4.2.3. Genomic modularity.[11] The final and most promising diversification mechanism we will discuss is presently called genomic modularity. The model was initially presented in 1999 at a meeting of the Baraminology Study Group, but since that time the model has undergone substantial revision. The model builds on the idea that God did not design genomes to be static and immutable, which fits well with our theological understanding of diversification. If God meant for His baramins to persist, He must have designed them to adapt. Similarly, if God meant for His organisms and their genomes to persist, He must have specifically designed the genome to adapt and change. Thus, instead of viewing a genome as something concrete, we ought to view it as a fluid structure, capable of sometimes dramatic change.

With this view of genomes in mind, we should expect to find features of the genome that are designed to diversify and change the genome but that do not participate directly in the biology of the individual organism. Wood called these features Altruistic Genetic Elements, or AGEs for short. AGEs are believed to act as mechanisms of genomic change, providing a means for reorganizing a single genome, for transferring information from one genome to another, and even for moving genetic material from one baramin to another. By doing so, AGEs alter the phenotype of the organisms upon which they act. We can think of genomes as designed to be interchangeable, or modular, hence the name *genomic modularity.*

Genomic modularity explains the characteristics of diversification better than any other proposed mechanism of diversification. The ability of genomes to be modular certainly requires a much higher level of design than a static, unchanging genome. The mechanistic specificity arises from the specificity inherent in the AGEs as they act on genomes. As we will see below, AGEs actually respond to environmental stress, providing a means of understanding both the stability of modern species and the dormancy of the mechanism, both of which may be due to the overall lack of biological and environmental stress. Though genomic modularity can explain all of these diversification features, much research is required to illuminate the specific details of how genomic modularity works.

We can observe genomic modularity by several methods, one of which is genomic sequencing of organisms that belong to the same baramin. Genome sequencing produces a precise map of all of the nucleotides of an organism's complete genome. By comparing genomes from co-baraminic organisms, we can detect changes that took place during the history of that baramin. Presently, these examples consist largely of bacterial genomes sequenced from strains of the same species, because the cost of sequencing eukaryotic genomes prevents a wide sampling of genome sequences from a single baramin. The comparative studies of bacterial genomes have revealed an amazing plasticity.

The bacterial genus *Chlamydia* consists of obligate intracellular parasites of eukaryotic cells and provides an excellent example of genomic modularity.[12] Presently, the genomes of four different *Chlamydia* strains have been fully sequenced, two strains of *C. trachomatis* and two of *C. pneumoniae*. The two *trachomatis* genomes show strong colinearity of homologous genes, except in a chromosomal region called the "plasticity zone."

The plasticity zone of *C. trachomatis* serovar D covers 23,000 nucleotides, but the same zone in *C. trachomatis* mouse pneumonitis strain Nigg covers more than twice that at 50,000 nucleotides. These regions contain a high concentration of genes associated with chlamydial pathogenicity. The size difference between them illustrates the rearrangements that have taken place after the divergence of the strains. The two *pneumoniae* strains are extremely similar, with very few insertions or deletions. One major difference was the presence of a 4,524-nucleotide bacteriophage integrated into the genome of *C. pneumoniae* strain AR39. Bacteriophages are ubiquitous features of bacterial genomes, frequently involved in the alteration of genes or transfer of genetic material between different bacteria. As such, we can classify bacteriophages as AGEs.

The differences observed in the chlamydial genomes illustrate the changes generated by genomic modularity, but because *Chlamydia* are so closely related, these changes probably occurred very recently. We can find changes directly associated with diversification by examining genomic differences in the various species of the grass holobaramin.[13] Because of their economic and agricultural importance as crops, many grass genomes have been mapped by using genetic and physical mapping techniques.[14]

The most obvious feature of grass genomes is their sizes. The smallest grass genome, that of rice, contains only 450 million nucleotides. The genome of maize is approximately three billion nucleotides, and the bread wheat genome contains around sixteen billion nucleotides. The size difference can be attributed to two phenomena: polyploidy and insertion sequences. Polyploidy occurs when a genome increases by some multiple of a complete

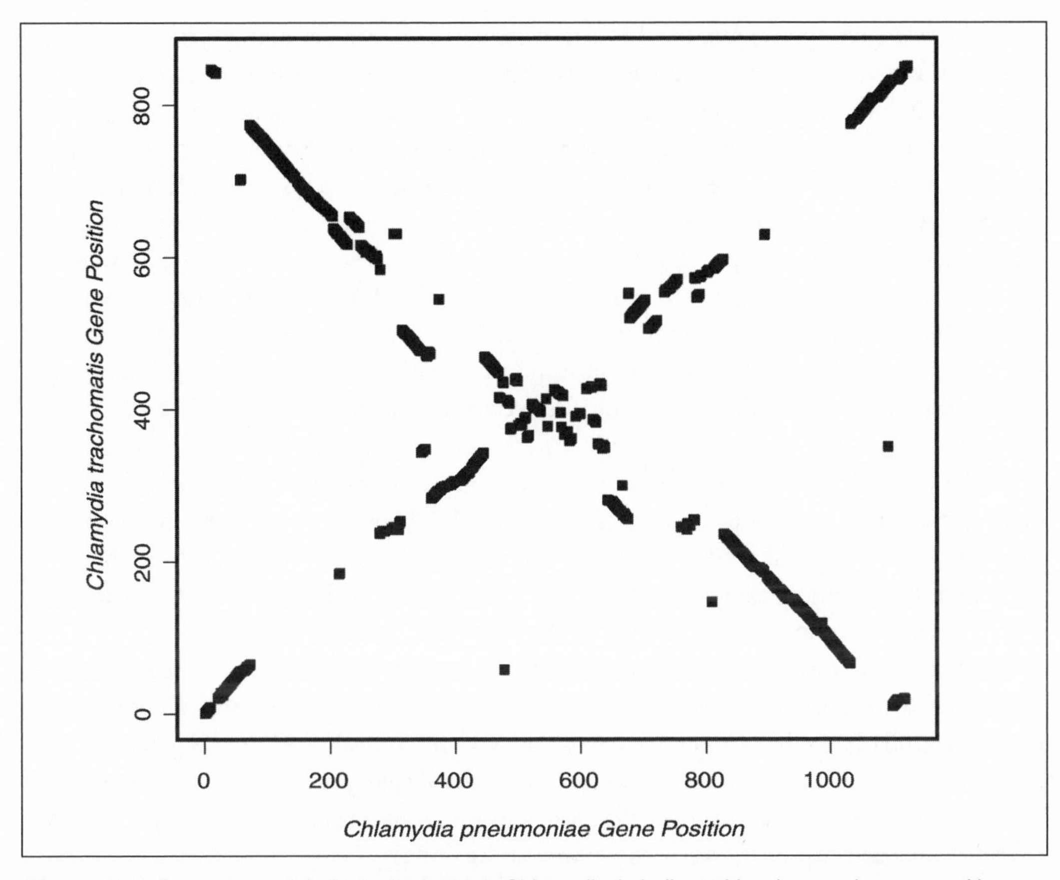

Figure 11.3. *Genomic modularity in the bacteria* Chlamydia *is indicated by changes in gene position. The genes of both genomes are numbered in order along their primary chromosomes. For each gene found in the* C. pneumoniae *genome, the position of the corresponding gene in the* C. trachomatis *genome is plotted along the Y axis. Lines of gene positions along the diagonal of the diagram indicate conserved gene order. Lines of gene positions perpendicular to the diagonal indicate large chromosomal inversions. The central region of the diagram (centering around gene 500 of* C. pneumoniae*) is extremely variable.*

set of chromosomes. The largest and most frequent insertion sequences found in grass genomes are retrotransposons. Because retrotransposons often influence gene expression by either disrupting the gene or altering its regulation, we can classify retrotransposons as AGEs.

In addition to the size differences, grass chromosomes also exhibit substantial differences in order and organization. The number of chromosomes also differs between different grass species. Although the chromosomal order of homologous genes is conserved in long regions, these regions are found on different chromosomes. For example, chromosome one of rice corresponds to regions of chromosomes two and three of sugarcane. Chromosome eight of sugarcane corresponds to chromosomes four and eleven of rice. During the diversification of the grass baramin, genomic modularity reorganized these blocks of genes to form different chromosomes in different species.

The retrotransposon AGEs found in the barley genome illustrate the response of AGEs and genomic modularity to environmental change and stress.[15] The *BARE*-1 retrotransposon occurs with an average frequency of fourteen thousand copies per barley genome, varying by species and variety. The first evidence we find that *BARE*-1 responds to the environment is that the frequency of *BARE*-1 correlates negatively with water availability. Barley plants grown under water stress have a higher copy number of *BARE*-1 than plants grown with sufficient water. Even better evidence of *BARE*-1 environmental response comes from an examination of wild barley plants from a canyon at Mt. Carmel, Israel. Although the plants in this canyon have the same climate, the amount of sunlight they receive varies by elevation and the direction of the slope they grow on. Not unexpectedly, *BARE*-1 copy number correlates with both elevation and slope face. As an AGE, *BARE*-1 has the ability to both reorganize the genome and alter gene expression in response to environmental changes. Considering the environmental stress associated with the recovery from the Flood, AGE activity must have been substantial during that period.

The concept of genomic modularity responding to environmental stress can explain diversification very well. During the extreme stress of the post-Flood recovery period, diversification would have occurred rapidly in response to the activity of the AGEs. As the stress decreased, AGE activity also would decrease, returning diversification to its dormant state. Many details of the mechanism are presently unknown. Although we have suggested that AGEs can exhibit a specificity of insertion (thereby inducing rapid change), many modern AGE elements induce detrimental mutations, possibly because of a degradation of the AGE function. Similarly, we have a poor understanding of the control of AGE and genomic modularity activity, particularly in response to environmental changes. In summary, although the genomic modularity model of diversification is more promising than other proposed diversification mechanisms, much research remains to be done.

11.5. Bacterial Antibiotic Resistance

As we have seen in this chapter, diversification took place during the post-Flood recovery period, but is presently not occurring. According to our best understanding, the diversification mechanism might be directly responsive to environmental stress. In present times, environmental stress is too slight to invoke the dramatic diversification that took place after the Flood, but some modern stresses can still induce important phenotypic changes in organisms through the mechanism of genomic modularity. As an example of modern diversification-like changes, we present the spread of antibiotic resistance among pathogenic bacteria.[16]

Alexander Fleming announced his discovery of the first antibiotic, penicillin, in 1929. He had discovered the substance accidentally after finding mold growing on a bacterial plate. The mold, *Penicillium notatum,* had killed the bacteria in a small circle around the mold growth. Fleming deduced that *Penicillium* must be producing a substance that killed the bacteria. Very soon after, many other anti-bacterial substances were discovered that were also harmless to people. Mass production of these new drugs came soon after, and the clinical use of antibiotics spread rapidly.

Antibiotics work by targeting proteins and processes in bacteria which are required for the bacteria to live or reproduce. For example, penicillin and substances like it (called ß-lactams) bind to a protein that is important for the synthesis of the bacterial cell wall, which is necessary for bacterial survival in the human body. Other antibiotics, such as tetracycline, disrupt protein synthesis in bacteria. Since proteins are central to metabolism, bacteria cannot live long when protein production is blocked. Some antibiotics, like quinolone and rifampicin, target DNA or RNA metabolism in bacteria, preventing their growth and reproduction.

As early as 1951, just twenty-two years after Fleming's announcement of penicillin, the first antibiotic-resistant bacteria were discovered in a clinical setting. Today, fifty years later, bacterial resistance to antibiotics is precipitating a health crisis. For every antibiotic produced by medical research, some strain of bacteria has developed a resistance mechanism, allowing that particular strain to survive despite the drug. The crisis is so severe that a few bacterial strains have become resistant to all antibiotics. Some strains of *Staphylococcus aureus* (which causes surgical infections) and *Mycobacterium tuberculosis* (which causes tuberculosis) can no longer be treated with any drugs.

Bacteria evade the effects of antibiotics by a variety of means. In some cases, simple mutations can bestow resistance to important antibiotics. For example, mutations in bacterial porin proteins prevent the import of ß-lactams, and mutations in RNA polymerase renders it impervious to inactivation by rifampicin. More frequently, resistant bacteria produce special enzymes that act directly on the antibiotic. Some of these enzymes modify the antibiotic and make it harmless to the bacterium. Other kinds of enzymes reduce the intracellular concentration of the antibiotic by actively pumping it out of the cell, reducing the damage it would do at high concentrations.

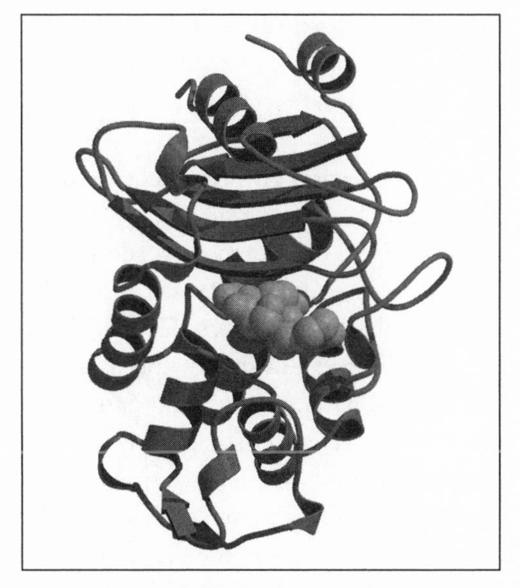

Figure 11.4. The molecular structure of the enzyme ß-lactamase from the bacterium Staphylococcus aureus. *The protein chain is shown as a series of ribbons. ß-Lactamases metabolize antibiotics such as penicillin, rendering them harmless to the bacterium. The penicillin is here represented by an inhibitor that binds to the protein in the same location as penicillin. The complexity of this enzyme implies that some antibiotic resistance mechanisms originate by design.*

In the majority of cases, antibiotic resistance spreads horizontally from one individual bacterium to another. Antibiotic resistance genes are found on plasmids and insertion elements, each of which can be moved about in the genome. Plasmids often encode complicated protein systems that allow bacteria to transfer genetic material from cell to cell. Thus, when a bacterial population with 1 percent antibiotic-resistant cells is challenged by an antibiotic, the resistant cells can quickly pass the resistance genes to cells that do not have them. As a result, bacterial antibiotic resistance can spread rapidly, as illustrated by the clinical history of resistant pathogens.

The history of vancomycin resistance provides a chilling example of the ability of antibiotic-resistance to spread through genomic modularity. In Europe, avoparcin is widely used as a livestock antibiotic. Bacteria that are resistant to avoparcin also show resistance to vancomycin, an antibiotic used in humans. After years of avoparcin use, vancomycin-resistant bacteria are common in Europeans who eat meat but not in European vegetarians or Americans, where avoparcin is not used. Somehow, avoparcin-resistance genes that originally developed in animal-inhabiting bacteria were transferred by a genomic modularity mechanism to bacteria that inhabit humans, possibly through ingestion.

The story of the rise and spread of antibiotic resistance fits well with our understanding of intrabaraminic diversification. The post-Flood environmental conditions made life difficult for most organisms, eliciting a diversification response that provided organisms with the ability to survive. In a similar way, the widespread use of antibiotics makes life difficult for most bacteria, eliciting a diversification response (genomic modularity) that provides bacteria with the ability to survive. Since God designed baramins to persist and survive, we should not be surprised that bacteria survive and thrive despite our best efforts to destroy them.

As we explained in chapter 11, pathogens derive from harmless organisms that become damaged by some effect of the Fall. Rather than viewing pathogens as something dangerous to be destroyed, medical researchers would be much more effective by treating the pathogen for the imperfection that causes it to be harmful. In this way, we can eliminate the pathogenicity and disease, and we can avoid inducing the natural diversification mechanisms that promote pathogen survival.

11.6. Chapter Summary

The question of species fixity and transformism remains contentious among a few creationists, but within baraminology, diversification is an inescapable conclusion. The pattern of diversification is a reconstruction of the historical pathway taken by a baramin through diversification. Although creationists widely accept phylogenetics within a baramin, this view may not always be correct. The widely accepted pattern of evolution is a simple tree structure, in which extant species appear as twigs or leaves and ancestral species appear as limbs, trunks, and roots. Under baraminology with mediated design stimulating the expression of predesigned characteristics, there is no compelling reason to believe that diversification is parsimonious.

We can infer the rate of diversification from the Bible and other ancient records. From these sources, we conclude that diversification must have been very rapid immediately after the Flood. The mechanism of diversification is an active area of research today. Based on the rate and pattern of diversification, we can infer four attributes of the mechanism: design, specificity, stability, and modern inactivity.

Numerous mechanisms have been proposed to explain diversification. Some creationists accept neodarwinian speciation, while others also advocate a mechanism called heterozygous fractionation. The most promising mechanism is genomic modularity, which can explain many features of genomes and may even have clinical applications.

Review Questions

1. Define reticulate evolution.
2. Define microevolution, speciation, and macroevolution.
3. Why should we use the term *intrabaraminic diversification*?
4. How does the rate of diversification differ from the rate of speciation?
5. How does the equid baramin demonstrate rapid diversification?
6. How does archaeology help us to understand the rate of diversification?
7. What analogy illustrates the pattern of diversification?
8. Why are organisms that are very similar assumed to be closely related? Is this necessarily correct? Why or why not?
9. Define convergent evolution.
10. List three mechanisms that creationists have proposed to explain diversification.
11. List the attributes of the diversification mechanism.
12. What diversification mechanism was advocated by Marsh?
13. Explain heterozygous fractionation.
14. Explain genomic modularity.
15. Explain why baraminologists should be careful equating degree of similarity with closeness of genealogy.
16. Does baraminology fit with the concept of common ancestry? Why or why not?
17. How does mediated design explain the independent appearance of new traits within a baramin?
18. What questionable assumptions are made by modern phylogenetic methods?
19. Define parsimony.
20. How does embryonic development help us to understand diversification?

For Further Discussion

1. Why are neodarwinian mechanisms of speciation widely accepted among creationists?
2. Explain how an evolutionist and a creationist might explain convergent evolution.
3. Why is speciation still a contentious subject to some creationists?
4. Explain and justify a method of treating a bacterial infection resistant to standard antibiotics.
5. Could we understand diversification from modern disaster recovery, for example after a major forest fire or a volcanic eruption? Why or why not?
6. Give five examples of evolutionary convergence.
7. If doctors had been more conservative in prescribing antibiotics, would antibiotic resistance have spread so rapidly among bacteria? Why or why not?
8. Explain how antibiotic resistance illustrates genomic modularity. Propose specific treatments for bacteria that focus on preventing harm to the host without killing the bacteria.

9. Given the mobility of modern people and the technological advances of modern medicine and biotechnology, could we observe another phase of diversification as organisms from so many different environments become mixed together?
10. Identify another example of genomic modularity in modern genomics.

CHAPTER 12

Baraminology and Biogeography

12.1. Introduction

Scientific challenges began to erode confidence in the straightforward creation account of Genesis during the seventeenth and eighteenth centuries. Among these challenges, biogeography is one of the earliest. Because they accepted a form of Linnaeus's species fixity, scientists had great difficulty explaining the geographical distribution of species. Ironically, creationist researchers have recently proposed powerful explanations of global biogeography that are consistent with the biblical record. As we will see in this chapter, these explanations go a long way toward resolving this early challenge to Genesis.

The science of biogeography is the study of the distribution of organisms on the earth. Biogeographers catalogue the occurrence of species and higher taxa in particular regions and propose hypotheses to explain how organisms came to be where they are. Beyond the mere description of species ranges, the explanation of species occurrence adds a definite historical dimension to biogeography. As such, modern biogeography is heavily influenced by evolution, the conventional model of historical biology. As we will see in this chapter, however, baraminology and the creation model of earth history can actually explain modern organismal distributions better than current evolution-based scenarios.

Biogeography of modern organisms deals with the distribution of creatures that has come about after the Flood. Because of its global scope, the Flood completely changed the original pattern of biogeography that existed from Creation. To understand the original geographical distribution of organisms, we must look to the fossil record and try to detect patterns that can be explained by geographical distributions. By doing so, we find that an ecological and biogeographical model of the fossil record actually explains fossil data very well.

We begin this chapter with a discussion of the geography of organisms from before the Flood, building on our best understanding of the fossil record. We will develop in more detail the hypotheses of pre-Flood organismal distributions that we previously alluded to in chapters 3 and 9. We will follow that with a brief review of biogeographic hypotheses through history, as explanations of the distribution of organisms following the Flood. We will then introduce a model of biogeography proposed by Wise and Croxton that builds directly on our best model of the Flood and baraminology. This new model can potentially explain most of the major patterns observed in the geographical distributions of organisms.

12.2. Biogeography Before the Flood

A common way to understand the fossil record is to view rock layers as a chronological progression of organisms over vast periods of time. From the simplest organisms in the lowest sediments to the complex mammals, angiosperms, and humans in the highest sediments, life seems to progress from simple to complex. Because this evolutionary perspective dominates modern academia, we often find it very difficult to imagine even the possibility of an alternative explanation, even though the alternative is very simple to understand. Instead of viewing the fossil record as a chronological, evolutionary development, we can view much of it as the successive inundation and burial of pre-Flood biological zones. This biogeographical view of paleontology explains fossils just as well as evolution.

Harold Clark was the first to propose ecological zonation as an explanation of the fossil record in his 1946 work, *The New Diluvialism*.[1] In his follow-up work, *Fossils, Flood, and Fire,* he developed his idea in more detail. Clark continued to promote his idea, and other creationists have since adopted and modified his concept.[2] Although Clark proposed that the progression of rock strata represented a progression from deep sea to high elevation, we here present a modified version of the same model which proposes biogeographic–rather than elevational–provinces.[3]

Before we begin discussing this model in detail, we must clarify terminology. The geologic column gives us many of the words we use to describe fossil taxa. Although originally formulated without a chronological or evolutionary bias, the geologic column has today become strongly associated with the extreme antiquity of the earth. Consequently, the major categories of rock strata are labeled with terms that indicate time. For example, Paleozoic means "ancient life," and Cenozoic means "new life." In place of these terms, we will use the original descriptive terms, Primary (= Paleozoic), Secondary (= Mesozoic), Tertiary (part of the Cenozoic), and Quaternary (part of the Cenozoic). According to the best creationist geology models, sub-Primary (pre-Cambrian) rocks were formed before the Flood, possibly at the formation of the dry land. All of the Primary and Secondary represent Flood sediments, while the Tertiary and Quarternary were deposited during the post-Flood period.

Beginning at the bottom of the column, we will work our way up and describe the ecological zones represented by each group of strata. Because the zones have little overlap between them, we infer that the organization of the pre-Flood organismal geography was much stricter than it is today. After the Flood, when all organisms were thrust together and then re-distributed, the geographical patterns that emerged display a great deal of gradation between ecological zones. In contrast, the pre-Flood zonation appears to have been much more well-defined, with very limited regions of ecological gradation.

Our first example of a pre-Flood ecological zone existed on the margins of the continents. Studies of the geology of the Death Valley region have revealed evidence of tremendous landslides associated with the onset of the Flood.[4] These landslides and other pre-Flood rocks from the southwestern United States contain evidence of a unique pre-Flood environment: a relatively shallow, epicontinental lagoon. This lagoon probably extended for hundreds of miles surrounding the pre-Flood coastlines, separated from the oceans by a shallow, hydrothermal environment probably formed during the uplift of the continents on Day 3. Closer to the shore, trilobites and associated animals thrived. Between these regions, where the water was deepest, soft-bodied, deep-sea animals typical of the Ediacara lived. At the outer margin, the hydrothermal ridge provided heat and shallow water for a community of bacteria. Typical of this region were giant stromatolites, bacterial mats growing on a mineralized substrate.

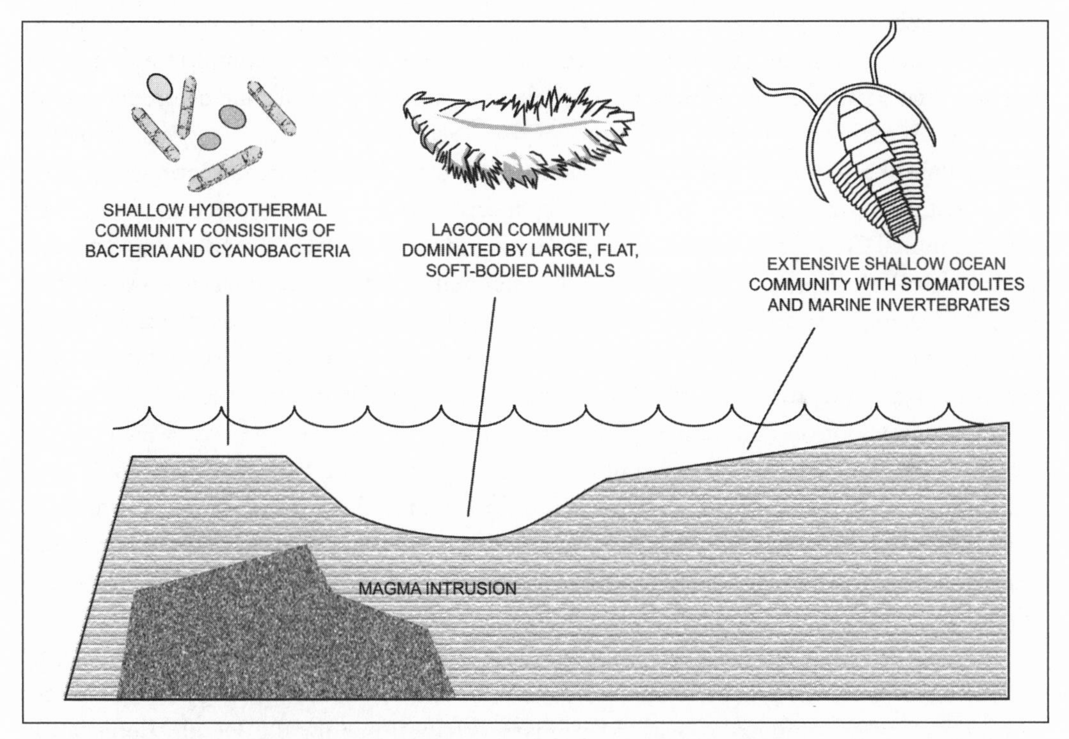

Figure 12.1. *Prior to the Flood, the margins of continents were relatively shallow, epicontinental lagoons, bounded by hydrothermal communities. The inundation of these lagoons produced fossil layers showing a gradation of bacteria from benthic organisms to shallow marine invertebrates.*

The boundary of the sub-Primary and Primary sediments generally represents the onset of the Flood, although there are some exceptions.[5] The unusual terrestrial plant and animal fossils of the Primary sediments represent the remains of a giant, floating forest, like a modern bog on a much grander scale.[6] This forest effectively acted as another "continent" in the pre-Flood world, providing a habitation for all manner of amphibians, insects, reptiles, and fish. The margins of the floating forest were composed primarily of rhyniophytes, small plants that were heavily dependent on water. The stability of the floating forest increased nearer the interior, allowing the growth of tree-like ferns, seed ferns, club mosses, horsetails, and cordaite gymnosperms. The Flood destroyed the forest from the outside in, burying the smaller, more water-dependent plants first and the larger, tree-like forms later. The main mass of the forest created the great coal seams of the Flood.

In the Secondary strata, we find what are arguably the fossils most popular with the general public, the dinosaurs. In the same strata, we also find many types of gymnosperms, which we infer served as the primary food source for the herbivorous dinosaurs. Because of the close geological association of dinosaurs with gymnosperms and with each other, we can conclude that a large region of the pre-Flood world was occupied nearly exclusively by these organisms. As the Flood waters progressed over the pre-Flood land mass, this region was one of the last to be destroyed, preserving these organisms near the top of the Flood deposits of the geologic column.

Strangely missing from the Flood-deposited strata were most of the mammals, birds, and angiosperms, and all evidence of human life. From their nearly complete absence, we can infer that these organisms probably occupied the same biological zone before the Flood. The few birds and mammals found in the Secondary strata suggests that the mammal/bird/human zone was probably adjacent to the dinosaur/gymnosperm zone. Even more exciting in this regard is the presence of morphological intermediate organisms, such as the mammal-like reptiles found in South Africa and the *Archaeopteryx* from a single deposit in Germany. If the border region between the mammal and dinosaur zones was a sharp ecological transition, it is possible that the intermediates long interpreted as evidence for evolution are actually evidence of an ecological transition. These transitional forms further support the geographic proximity of the pre-Flood dinosaur and mammal zones.

If mammals, birds, and humans lived together in an ecological zone, why were they not preserved like the other pre-Flood zones? One answer might be that mammals, birds, and humans make poor candidates for fossilization in the first place. This answer, however, is much less satisfying when we consider the amazing preservation of delicate mammal and bird fossils in the Secondary (such as *Archaeopteryx*). A more satisfying solution to the whereabouts of the pre-Flood mammals relates to the mechanism of the Flood itself.

When the Flood began, large portions of the crust that formed the ocean floor began to slide underneath the crust that formed the continents. Creationist geologists believe that this process, called subduction, was the primary driving force for the Flood.[7] Because the subduction occurred so quickly (destroying thousands of kilometers of ocean crust in months), the subduction zone would have been a site of violent geological upheaval. If the mammal zone bordered on one of these subduction zones, it is possible that all remnants of the pre-Flood people (and mammals and birds) were annihilated in the subduction zone.

Alternatively, the spring which fed the river of Eden might have been centered on the "fountains of the great deep" (Gen. 7:11; 8:2). If humans, mammals, and angiosperms lived near this region (which they probably did), the breakup of the fountains during the Flood could have easily obliterated any trace of these organisms. More research on both the fossil record and the Flood itself will help us formulate a more precise answer to this intriguing mystery.

From this brief review, we can see how the fossil record can provide important insight into the biogeographic conditions before the Flood. Rather than simply discarding fossils as evidence of evolution, a careful reinterpretation places them in their proper context as evidence of pre-Flood biogeography. Further research in paleontology will undoubtedly continue to illuminate important and exciting information about the world that perished.

12.3. A History of the Science of Biogeography

After the Flood, creatures repopulated the earth from centers of survival as we described in chapter 3. Terrestrial animals migrated from region of Ararat, in what are now the mountains of Turkey, Armenia, and Iran. All organisms not present on the ark would have recovered from any number of centers of survival. Plants could reestablish themselves vegetatively as well as through seeds, and the growing animal population would soon have sufficient food to sustain it. Modern biogeographers study the outcome of this repopulation and redistribution of organisms. Consequently, the history of biogeography itself begins with early ideas of harmonizing observed organismal ranges with the biblical Flood narrative.[8]

With a little imagination, we can easily see how appealing the early model of biogeography was. Back before the age of exploration when biologists still believed in a form of species fixity, species ranges were explained as simple dispersion from the ark. Because species could not change and because all terrestrial species were present on the ark, the animals leaving the ark must have simply wandered the earth until they arrived at a region of suitable climate. Before the tales of global exploration became well known, this explanation of biogeography was very reasonable.

The many voyages of the European explorers during the sixteenth and seventeenth centuries produced much new data that challenged this early, simple view. First, explorers noted that regions with similar environmental conditions (such as the various tropical areas) often contained very different species, an unexpected result. If climate was the only determining factor in the distribution of organisms, we would expect that identical climate zones would be settled by identical species wandering from the ark. Second, the very existence of so many isolated land masses (such as Australia, the Americas, and the many Pacific islands) seemed to prevent wandering terrestrial animals from colonizing these regions. Since these islands were inhabited by terrestrial animals, they probably did not arrive there by simple migration.

At this point, scientists rejected the dispersion model. Instead of modifying their view of species fixity, most modified their view of Scripture. To explain the good fit between organisms and their environments and the disjunct or isolated distributions, biologists suggested that God re-created species after any number of catastrophes had wiped them out. Since species were incapable of changing–and they certainly could not have traversed the wide oceans to get to isolated islands–God must have created them in the environment where we now find them, presumably perfectly adapted for their particular conditions of existence.

Although this theory does explain how organisms get to remote areas of the earth, it does so arbitrarily. Indeed, we could explain *any* anomalous or unusual observation by appealing to God's direct intervention. As more data accumulated about organisms and their environments, this view of biogeography also fell out of favor. Primarily, the re-creation model presumed that God perfectly designed organisms to live in the precise environment in which they are found. In contrast, scientists quickly discovered that some organisms, such as rats, could thrive in any number of exotic environments. Clearly, the environment they were created in was not the only perfect fit.

With the increase in popularity of species transformism in the eighteenth and nineteenth centuries, biogeography acquired a powerful means of explaining similar species in adjacent regions. Under the two preceding creation models built on species fixity, similarity and dissimilarity of organisms was relegated to the mysterious will of God. Under transformism, dispersion again became a popular biogeographical explanation, except that dispersion of higher taxonomic categories became interpreted as the geographical evolution of a particular lineage.

This model of biogeography became immediately problematic because it predicts a random distribution of centers of dispersal, but the earth does not have such a random distribution. Instead, the earth exhibits regions of high **endemism**, where unique species are found in only one region. Because evolution happens all over the planet, there is no particular reason to expect that one region would be preferred over another for the origin of a new organismal lineage. Consequently, regions of high endemism are not expected under evolution. At the same time, many taxa also show a highly disjunct distribution, with closely related species found in very distant regions, especially across oceans. Dispersion of evolving lineages explain these disjunctions very poorly.

In the 1960s, plate tectonics became widely accepted in geology, introducing a new explanatory tool to biogeography: **vicariance**. According to the plate tectonic theory, the present continents were once joined together in various configurations that allowed terrestrial organisms to disperse to regions that today are separated by water. Thus, organisms dispersed vicariously through the movement of land masses. When plate tectonics was first proposed, disjunct species distributions were used as evidence for the past connection of continents.

Today, vicariance biogeography has also fallen out of favor. Geologists now believe that continental drift took many millions of years. Ironically, the disjunct species distributions initially used as evidence of past continental connections are now recognized as being far too young to be explained by continental drift. Furthermore, there are several disjunctions that cannot be explained by continental drift because the continents in question were never joined in the way that the distribution implies. For example, many species and genera have disjunct distributions across the Pacific Ocean, but plate tectonics does not postulate a direct connection of the lands across the Pacific. Consequently, these disjunctions are not explained by vicariance.

Today, in the absence of a guiding model, biogeography is largely an *ad hoc* exercise. Organismal distributions are explained by the occasional and accidental dispersal of organisms by wind, birds, people, or even floating on an isolated log across oceans. These explanations are very random and arbitrary, and consequently almost impossible to test. Remembering what we have learned of baraminology in this book, we can see why the historical explanations have failed. Species fixity simply does not adequately describe reality, and subsequently all models built on species fixity will suffer similar problems. At the same time, the long time periods associated with evolution prevent a ready explanation of similar organisms found in isolated regions. By accepting the diversification allowed under baraminology as well as the rapidly changing environment predicted by modern Flood models, we can begin to see a possible explanation of post-Flood biogeography.

12.4. Biogeography after the Flood

Until only recently, creationists have largely ignored the science of biogeography, but a new biogeography model goes a long way toward rectifying that neglect. Wise and Croxton's post-Flood rafting model synthesizes theories of Flood geology, post-Flood catastrophism, and intrabaraminic diversification. According to their model, the debris produced by the destruction of the pre-Flood forests formed large mats during the Flood. These mats remained very stable well after the Flood and provided a means for organisms to disperse from Ararat by rafting, without the necessity of remaining on land or following coastlines. As we will see, this proposal explains a great deal of otherwise-enigmatic biogeographic data. Like all new theories, the post-Flood rafting hypothesis will continue to be developed and refined in the future.

Far from being a construction of pure fancy, large debris mats have been observed in our modern world. On the morning of May 18, 1981, Mt. St. Helens erupted with an enormous steam blast and landslide. The force of the explosion was directed horizontally north, leveling 150 square miles of old-growth forest in six minutes. Approximately one million logs ended up floating in Spirit Lake, forming a mat that covered two square miles. Although many of the logs have subsequently sunk into the muck at the bottom of the lake, a sizeable log mat still floats there today.[9]

As we examine rock strata deposited after the Flood, we find much evidence of severe volcanic activity, implying that the Flood itself must have been associated with even greater volcanism. For example, in Yellowstone National Park, Specimen Ridge testifies to a massive floating log mat that resulted from post-Flood volcanic eruptions. Specimen Ridge contains twenty-seven layers of logs deposited in a manner similar to the logs in Spirit Lake and covering an astonishing forty square miles.[10] As we described in chapter 3, the destruction of the pre-Flood forests by the combined action of volcanos and advancing water must have left an enormous amount of debris floating in the water. Based on the size of modern coal beds, we can estimate the size of these log mats to be hundreds of square miles.

According to Wise and Croxton's model, the size and density of these debris mats provided two important biological services. First, the mats provided a means of plant survival during the Flood, as described in chapter 3. Second and more important for our present discussion, these mats could also provide a stable platform on which animals could be distributed after the Flood. Because the mats were the centers of recovery for many plant species, they would have attracted herbivores, which in turn would attract carnivores. Inevitably, animals would have wandered onto the mats themselves, which subsequently transported them hundreds of miles along the ocean currents.

As the logs and debris forming the mats became waterlogged, the mats broke up and sank. Based on observations of the log mat at Spirit Lake, the stability of the post-Flood log mats would decay exponentially. As a result, rafts in the decades immediately following the

Figure 12.2. *A log mat of a million logs floats in Spirit Lake shortly after the eruption of Mt. St. Helens. Many of these logs remain floating to this day.*

Flood would have been very stable and capable of carrying large animals. Very soon thereafter, though, the mats would become increasingly incapable of carrying terrestrial animals.

Rafting of organisms on floating debris mats sounds almost *ad hoc,* but in fact, the hypothesis explains a tremendous amount of otherwise anomalous data. First of all, the areas of endemism, discussed above as a challenge to evolutionary biogeography, are readily explained by rafting when we realize that the areas all correspond to regions where rafts would have landed after the Flood. Second, ranges of genera or even species that are interrupted by oceans can also be explained when we view the ocean as a quick pathway from one region to another instead of an impassable barrier.

In addition to these general features of biogeography, many specific peculiarities can be explained by post-Flood rafting. The fossil record of the horse baramin contains an important anomaly. Although most of the early browsing horses are known from North America, *Hyracotherium* is also found in Europe.[11] Horses do not appear in Asia until late in the post-Flood period. If we had to explain the dispersal of the horse baramin by following contiguous land masses, we should expect some evidence of the passage of *Hyracotherium* through Asia as they traveled to the New World. With giant, stable debris rafts, the Atlantic ocean ceases to be a dispersal barrier and becomes a dispersal opportunity. The disjunction in the *Hyracotherium* distribution is expected as the horse baramin traverses the Atlantic on its way to the New World.

Figure 12.3. *The relatively minor eruption of Mt. St. Helens leveled 150 square miles of trees like these. The scientists in the bottom right of the photograph give a sense of scale to the devastation.*

Rafting can also help explain a classic textbook example of vicariance biogeography within a creationist view of history.[12] The various flightless birds of the southern hemisphere called ratites include the South American rhea, the African ostrich, the New Zealand kiwi, the emu and cassowary of Australia and New Guinea, and the extinct elephant bird of Madagascar and moa of New Zealand. These birds have long challenged creationist understanding of biogeography. In particular, the presence of the environmentally sensitive kiwi in New Zealand is thought to be particularly problematic to creationists. In contrast, the breakup of the southern continent Gondwanaland is proposed by many evolutionists to explain the modern distribution of ratites. As Gondwanaland fractionated into South America, Africa, Australia, and New Zealand, the flightless ratites merely went along for the ride.

Although no ratite baraminology study has yet been published, Wise and Croxton's rafting hypothesis provides some important insight into ratite distribution. A current that circles the globe around the Antarctic provides a direct connection between South America, Africa, Madagascar, Australia, and New Zealand via post-Flood debris rafts. A second current from the west coast of Africa to the east coast of South America provides a second dispersal route for ratites. If all modern ratites belong to the same baramin, we might hypothesize that dispersal and diversification took place in parallel as the ancestral ratites rafted along the circum-Antarctic current. Even if modern ratites consist of more than one baramin, rafting still provides an excellent mechanism for transporting flightless birds to isolated islands.

12.4.1. Island Biogeography

Combined with an appreciation for intrabaraminic diversification as discussed in the previous chapter, rafting can provide a powerful mechanism for understanding island endemism. Historically, island endemism played an important role in the intellectual development of modern biology. The observation of unique species on isolated islands had a great impression on Darwin as he began formulating his ideas of evolution. When we consider rafting and diversification together, modern creationists can predict the types of organisms that will be found on an island, potentially in greater detail than evolutionists could.

Because we know that diversification effectively ceased soon after the Flood, we can use degree of biological divergence to approximate (very roughly) the date that a species or group of species originated. Islands that emerged during or immediately after the Flood would have been colonized by organisms that were still diversifying and floating on the largest debris mats. Consequently, we should find that the oldest islands contain on average larger organisms that are more divergent from their mainland counterparts. Islands that emerged later in the post-Flood period would have been colonized by organisms that might have been nearly finished diversifying and that used rafts that were much less stable than the earlier post-Flood rafts. As a result, we should find that young islands contain on average smaller organisms that are less divergent from their mainland counterparts. As examples of these types of islands, we will discuss Australia, the Galápagos, and the recent islands Surtsey and Anak Krakatau.

12.4.1.1. Australia. As the smallest continent, Australia would have numerous starting and stopping points for post-Flood rafting. We can see the effects of this today in the organismal affinities with such distant lands as Madagascar and India.[13] By far, the most famous biogeographic feature of Australia are the marsupials. Aside from the possum, all living marsupials are endemic to Australia and surrounding islands. Not surprisingly, fossil forms of Australian marsupials are also found exclusively on Australia.[14] Although the discontinuity with other mammals is obvious, the presence of marsupials on Australia poses two fascinating questions for the creationist. First, how did Australian marsupials make it to Australia without a land bridge? Second, why are they found only in Australia?

To answer the first question of how marsupials got to Australia in the first place, we can cite the large, stable debris rafts that existed after the Flood. When we remember that Noah probably took smaller varieties of each baramin with him on the ark (see chap. 3), the presence of large animals on any remote island becomes very reasonable. The smaller members of each baramin rafted on the post-Flood rafts and later diversified into much larger species in their new environments.

Figure 12.4. *Marsupials like this wallaby (*Macropus*) are only indigenous to Australia and surrounding islands. Although widely believed to be excellent evidence for evolution, the post-Flood rafting model could provide an effective explanation of modern marsupial distributions.*

To answer the second question of why marsupials are found in Australia, we must turn to the post-Flood fossil record for clues. The first thing we notice from the marsupial fossil record is their ubiquity. After the Flood, various marsupial baramins appear everywhere, including Africa, Asia, Europe, and even Antarctica. Their widespread occurrence testifies to their successful rafting ability, but their modern absence from the same regions indicates that they were unable to establish themselves as the dominant animals in their ecosystems.

A second observation from the marsupial fossil record helps explain why marsupials did not become established anywhere else on the modern earth.[15] After the Flood, we find two major concentrations of marsupial fossils in Australia and South America. During most of the post-Flood recovery, the North and South American continents were not connected by the isthmus of Panama. Instead, South America was an island continent much like Australia. During its isolation from other land masses, South America came to be dominated by marsupials and giant, flightless, carnivorous birds. The post-Flood fossil record shows a largely placental mammalian fauna in North America. When the isthmus of Panama connected the two continents, many placentals from North America invaded

South America, but comparatively few South American marsupials invaded North America. As a result, nearly all South American marsupials went extinct.

The post-Flood history of Australia resembles that of South America in many important respects save one: placental mammals never invaded Australia in any great numbers. Based on these important observations, we can infer a general narrative that helps to explain Australian biogeography. After the Flood, marsupials apparently dispersed more rapidly than placentals, and as long as the period of rapid diversification lasted, the marsupials were able to adapt to ecological stresses induced by placental invaders. Once the main period of diversification stopped, marsupial baramins could not react fast enough to adapt to placental invasion. Thus, when South America joined with North America at the close of the diversification period, the placentals drove the marsupials to extinction. Australia never experienced a strong placental invasion, probably because the debris rafts had become too unstable by the time placental mammals came to a launch point. Consequently, marsupials still thrive in Australia today.

12.4.1.2. Galápagos. The famous Galápagos islands are home to important examples of adaptation that aided Darwin in formulating his idea of natural selection. Endemic species on the Galápagos archipelago include a flightless cormorant, marine iguana, giant tortoises, and numerous species of finch (known as Darwin's finches). Surprisingly, when considered within the framework of baraminology and post-Flood rafting, the Galápagos actually provides an excellent example of post-Flood recovery in the creation model.

The Galápagos archipelago lie very close to the conjunction of two tectonic plates, right on the equator.[16] The islands themselves are volcanic, like the much larger islands of Hawaii. Although there exists some debate about the exact age of the islands, they definitely formed very late after the Flood, possibly coincident with the joining of North and South America. Today, an equatorial ocean current flows from mainland South America to the archipelago six hundred miles away.

With the dominant ocean current coming from South America, we should expect that the organisms on the Galápagos would probably be from the same baramins as on the mainland. The lateness of the islands' formation should have two effects. First, organisms probably colonized the islands on the tail end of the diversification period. Consequently, they should not be substantially different from members of the same baramin in South and Central America. Second, the post-Flood debris rafts probably broke into smaller, less stable pieces by the time the Galápagos formed. As a result, we should expect to find few large animals on the islands.

Generally, these predictions are evident when we look at the flora and fauna of the Galápagos. Because the Galápagos were colonized so late after the Flood, the species we find there are closely related to similar species both in the islands and on the mainland. For example, three finch species of the genus *Geospiza* interbreed on Isla Daphne Major, indicating that they belong to the same baramin.[17] The *Geospiza* hybrids are both viable and fertile, implying that the parental species are closely related. The famed marine iguana *Amblyrhynchus cristatus* occasionally hybridizes with the Galápagos land iguana *Conolophus subcristatus*.[18] Despite the fact that no other diving iguanas are known in the world, both Galápagos iguana species belong to the same baramin.

Figure 12.5. The Galápagos Archipelago formed from a series of volcanic eruptions after the Flood. Today they are home to a number of unusual species found nowhere else in the world. These species are all very similar to other species of the same baramin found on mainland South and Central America.

By examining similarities between Galápagos and mainland fauna, we definitely find that the organisms on the archipelago appear to have arrived from the mainland of South America. For example, according to a study of mitochondrial DNA, Darwin's finches are most similar to the dull-colored grassquit (*Tiaris obscura*), which lives in both Central and South America.[19] The mitochondrial DNA of Galápagos tortoises are most similar to the tortoises of South America.[20] As expected, the organisms of the islands are closely allied with the mainland to the east.

The enormous size of the Galápagos tortoises seems to contradict the expectation that only small animals should inhabit the Galápagos. We can find clues to this apparent anomaly by examining a wider variety of evidence concerning post-Flood tortoise dispersal and diversification. As mentioned above, mitochondrial DNA analysis has shown a close relationship between the Galápagos tortoises and mainland tortoises.[21] The closest relationship is to the Chaco tortoise *Geochelone chilensis,* the smallest of the three South American *Geochelone* species. The average shell length is 43 cm, less than half of the length of a typical Galápagos specimen. We might interpret this relationship as indicating a small species ancestral to the Galápagos tortoises initially colonized the islands after the Flood.

Alternatively, we might argue that larger species of tortoises raft better than smaller ones since they would be less likely to be swept into the oceans during the six-hundred-mile journey. Fossilized giant tortoise remains found in South America might support the view that the tortoises that colonized the Galápagos were already large. By analogy, we can find evidence in other remote islands of large tortoises. Giant tortoises are also known from a number of islands in the Indian Ocean. Today, an endemic population is found only on the Aldabran islands. The presence of other giant tortoises on islands suggests that giant tortoises may be more suited to rafting than smaller ones. More research will undoubtedly help to resolve this fascinating question.

Figure 12.6. *The largest tortoises in the world, such as the Galápagos tortoise (*Geochelone nigra, *bottom) and the Aldabran tortoise (*G. gigantea, *top), are found on remote islands. Post-Flood debris rafting provides an effective explanation for the presence of these tortoises on isolated islands.*

12.4.1.3. Surtsey and Anak Krakatau.[22] Our understanding of the history of creation leads us to conclude that rapid diversification and rapid transoceanic dispersal do not happen today like they did immediately after the Flood. As a result, volcanic islands that emerged very recently should display a small number of species that arrived by random processes from nearby land masses. Two volcanic islands that formed in the twentieth century fit this description, and they allow us to observe modern dispersal and invasion of barren habitats. A 1930 volcanic eruption in Indonesia annihilated the entire island of Anak Krakatau, providing a means of understanding dispersal in a tropical climate. The island of Surtsey just south of Iceland was formed in 1963, allowing observation of dispersal in colder climates.

Plants are the most diverse organisms on these islands. Because of its warm climate and proximity to other islands, Anak Krakatau has a higher rate of species invasion than Surtsey. Surtsey is populated by several dozen plant species, twenty to thirty invertebrate species, and six species of sea birds. In contrast, the species concentration on Anak Krakatau is much higher. Anak Krakatau is home to twenty-four species of land bird, in addition to bats and reptiles. As expected, the species colonizing these islands are the same species that may be found on nearby land masses. For example, 72 percent of Surtsey's species dispersed from the closest islands, but only 21 percent are believed to have come from the mainland of Iceland, twenty-seven miles away.

These newly formed islands hold important lessons for us as baraminologists. First, we can see that the species invading these new lands arrive by chance and populate the new lands slowly. In contrast, the dispersal after the Flood appears to have taken place very quickly, supporting the idea that a different mechanism, such as rafting, was operating at that time. Second, we also see that the species arriving do not diversify, further supporting a limited time period of diversification that followed immediately after the Flood.

12.5. Chapter Summary

Biogeography was one of the earliest challenges to erode confidence in the straightforward creation account of Genesis. Biogeography is the study of the distribution of organisms on the earth. Biogeography of modern organisms deals with the distribution of creatures that has happened after the Flood.

Biogeography from before the Flood may be inferred from the fossil record. A common way to understand the fossil record is to view rock layers as a chronological progression of organisms over vast periods of time. Life seems to progress from simple to complex. Harold Clark was the first to propose ecological zonation as an explanation of this fossil trend. Clark proposed that the progression of rock strata represented a progression from deep sea to high elevation. Today, this model has been extensively refined by modern creationist research.

Before the Flood, life was organized into several strictly defined ecological zones. One pre-Flood ecological zone existed on the margins of the continents, comprised of a relatively shallow lagoon bounded by a hydrothermal community. The Primary sediments also provide evidence of a colossal floating forest that may have covered hundreds of miles. In the Secondary strata, we find the dinosaurs and gymnosperms, which we infer served as the primary food source for the herbivorous dinosaurs. Strangely missing from the Flood-deposited strata are most of the mammals, birds, and angiosperms, and all evidence of human life. From their nearly complete absence, we can infer that these organisms probably occupied the same biological zone before the Flood.

The history of the modern science of biogeography begins with early ideas of harmonizing observed organismal ranges with the biblical Flood narrative. Before the age of exploration when biologists still believed in a form of species fixity, species ranges were explained as simple dispersion from the ark. The many voyages of the European explorers during the sixteenth and seventeenth centuries produced much new data that challenged this early, simple view. As a result, scientists rejected the dispersion model. Instead of modifying their view of species fixity, most modified their view of Scripture.

In the 1960s, plate tectonics became widely accepted in geology, introducing a new explanatory tool to biogeography: vicariance. According to the plate tectonic theory, the present continents were once joined together in various configurations that allowed terrestrial organisms to disperse to regions that today are separated by water. Today, vicariance biogeography has also fallen out of favor.

Until only recently, creationists have largely ignored the science of biogeography, but a new biogeography model goes a long way toward rectifying that neglect. Rafting can provide a powerful mechanism for understanding island endemism. Australia, the Galápagos, and the recent islands Surtsey and Anak Krakatau all fit within this new creationist model of biogeography.

Review Questions

1. Define biogeography.
2. Why is the Flood a pivotal event for biogeography?
3. How do we derive information about pre-Flood biogeography?
4. Define subduction.
5. Define vicariance.
6. How did organisms most likely disperse after the Flood?
7. How did the history of biogeography challenge the biblical Flood story?
8. What model was first proposed by Harold Clark?
9. How can fossils help us understand biogeography from before the Flood?
10. Define ecological zonation.
11. Explain the correlation between fossil layers and pre-Flood biogeography.
12. Why was the dispersion model abandoned under species fixity?
13. Did conventional plate tectonics provide a solution to biogeography? Why or why not?
14. Describe Wise and Croxton's floating debris mat theory.
15. What does Mt. St. Helens have to do with post-Flood biogeography?
16. How does rafting help us to explain the distribution of the flightless ratite birds?
17. How does rafting help us to explain endemic Australian marsupials?
18. How does rafting help us to understand the biogeography of the Galápagos islands?
19. Give two evidences that the flora and fauna of the Galápagos originated from mainland South America.
20. Where did the organisms found on Surtsey originate?

For Further Discussion

1. Research the biogeography of Hawaii and propose a post-Flood rafting hypothesis to explain it.
2. Given the relationship between pre-Flood geography and the fossil record, propose and justify three ways something might not be fossilized in the Flood.
3. Could rapid diversification and transoceanic dispersal ever happen again like it did immediately after the Flood? Why or why not?

Epilogue

Science is a peculiar thing. Everyone (and I mean *everyone*) practices science in one form or another, but very few people know it. Far too many people look at science as a body of knowledge carefully guarded by the precious few who can actually understand it. This is probably caused and perpetuated by horrendous definitions of science in high school and college textbooks. In reality, science is merely one way of gaining knowledge about the world around us. Scientists are very pragmatic. If a theory matches reality, a scientist will accept it. If the theory does not match reality very well, a scientist might reject it or perhaps hold on to it until a better theory presents itself. In the end, whatever works wins.

Science sounds pretty unglamorous when I describe it like that, but the very pragmatism of science is what makes it fun. There will always be anomalies and mysteries that confound even the most brilliant among us, partly because of human imperfection and partly because of the incomprehensibility of God and His creation. Because of that enduring mystery, a scientist always has something to do. Some might view this as ultimately frustrating, but I look at it as quite satisfying. I could understand frustration if God did not want to be discovered, but because God reveals Himself to us, I can know truth about Him. This gives me a basis for believing that some scientific models will more closely approximate truth than others. Thus, I do tend to believe that good science will progress toward truth (although it does not necessarily do so and it will probably never arrive *at* truth on this side of heaven). By building up baraminology on what has come before me, I contribute to that journey toward the truth. I believe that's quite a wonderful occupation, constantly searching and thinking to discover more about God and His works. Solomon recognized this when he wrote, "It is the glory of God to conceal a thing: but the honour of kings is to search out a matter" (Prov. 25:2).

As I muse about science, I see that my book (unsatisfying though I think it is) fits the bill. I picked up where others have left off, filling in some of the more gaping holes of previous theories. At the same time, the present book also has a lot of holes in it, areas of biology that don't yet fit quite right. But those holes will give my students plenty to do long after I'm gone. In the end, I'm certain that some day someone will come up with a model that fits biology and the Bible fairly well.

I would like to conclude the book with simple words of encouragement. You've read about the current state of creation biology. I've given you tools and advice. Now we need you out there working on those holes, not just sitting and reading this book. We *need* all capable Christian biologists to contribute to the creation model in whatever way they can. Put the book down, get out your microscopes, butterfly nets, computers or whatever you use, and get busy for the glory of God!

Oh, and have fun too.

Selected Bibliography

Cavanaugh, D. P., and T. C. Wood. A Baraminological Analysis of the tribe Heliantheae *sensu lato* (Asteraceae) using Analysis of Pattern (ANOPA). *Occasional Papers of the BSG* 1 (2002): 1–11.

Frair, W. Baraminology, Classification of created organisms. *Creation Research Society Quarterly* 37(2000): 82–91.

Gray, A. P. *Mammalian Hybrids*. 1st ed. Farnham Royal, Bucks, England: Commonwealth Agricultural Bureau, 1954.

———. *Bird Hybrids: A Check-List with Bibliography*. Farnham Royal, Bucks, England: Commonwealth Agricultural Bureaux, 1958.

Helder, M. J., editor. *Discontinuity: Understanding Biology in the Light of Creation*. Cedarville University: Baraminology Study Group, 2001.

Jones, A. J. How many animals in the Ark? *Creation Research Society Quarterly* 10 (1973): 102–8.

Marsh, F. L. *Fundamental Biology*. Lincoln, Neb.: Published by the author, 1941.

———. *Evolution, Creation, and Science*. 1st ed. Washington, D.C.: Review and Herald Publishing Association, 1944.

———. *Evolution, Creation, and Science*. 2nd ed. Washington, D.C.: Review and Herald Publishing Association, 1947.

———. *Studies in Creationism*. Washington, D.C.: Review and Herald Publishing, 1950.

———. *Variation and Fixity in Nature*. Omaha, Neb.: Pacific Press Publishing Association, 1976.

Reynolds, J. M. Intelligent design and discontinuity: Platonic metaphysics as a motivating worldview for discontinuous biological structures. *Occasional Papers of the BSG* 2, in press.

Robinson, D. A. A mitochondrial DNA analysis of the Testudine apobaramin. *Creation Research Society Quarterly* 33 (1997): 262–72.

Robinson, D. A., and D. P. Cavanaugh. A quantitative approach to baraminology with examples from catarrhine primates. *Creation Research Society Quarterly* 34 (1998a): 196–208.

Robinson, D. A., and D. P. Cavanaugh. Evidence for a holobaraminic origin of the cats. *Creation Research Society Quarterly* 35 (1998b): 2–14.

Robinson, D. A., and P. J. Williams, editors. *Baraminology '99: Creation Biology for the 21st Century.* Liberty University: Baraminology Study Group, 1999.

Scherer, S., editor. *Typen des Lebens.* Berlin: Pascal-Verlag, 1993.

———. Basic Types of Life: Evidence of Design from Taxonomy? *Mere Creation.* Editor W. A. Dembski. Downers Grove, Ill.: InterVarsity Press, 1993.

VanGemeren, W. A., editor. *New International Dictionary of Old Testament Theology and Exegesis* (NIDOTTE in endnotes). Grand Rapids, Mich.: Zondervan, 1997.

Westermann, C. *Genesis 1–11.* Translator J. J. Scullion. Minneapolis: Fortress Press, 1994.

Williams, P. J. What does *min* mean? *Creation Ex Nihilo Technical Journal* 11 (1997): 344–52.

Wise, K. P. Baraminology: A young-earth creation biosystematic method. *Proceedings of the Second International Conference on Creationism,* editors R. E. Walsh, and C. L. Brooks, vol. 2, 345–58. Pittsburgh: Creation Science Fellowship, 1990.

———. Practical Baraminology. *Creation Ex Nihilo Technical Journal* 6 (1992): 122–37.

———. Is life singularly nested or not? *Proceedings of the Fourth International Conference on Creationism,* editor R. E. Walsh. Pittsburgh: Creation Science Fellowship, 1998.

Wise, K. P., and M. Croxton. Rafting: A post-Flood biogeographic dispersal mechanism. *Proceedings of the Fifth International Conference on Creationism,* editor R. E. Walsh. Pittsburgh: Creation Science Fellowship, 2003.

Wood, T. C. A baraminology tutorial with examples from the grasses (Poaceae). *TJ* 16 (2002a): 15–25.

———. The AGEing process: Rapid post-Flood intrabaraminic diversification caused by Altruistic Genetic Elements (AGEs). *Origins* 53, in press.

———. Perspectives on AGEing, a young-earth creation diversification model. *Proceedings of the Fifth International Conference on Creationism,* editor R. E. Walsh. Pittsburgh: Creation Science Fellowship, 2003.

Wood, T. C., and D. P. Cavanaugh. A baraminological analysis of subtribe Flaveriinae (Asteraceae: Helenieae) and the origin of biological complexity. *Origins* 52 (2001): 7–27.

Wood, T. C., D. P. Cavanaugh, and K. P. Wise. Baraminological Studies of the Fossil Equidae. *Proceedings of the Fifth International Conference on Creationism,* editor R. E. Walsh. Pittsburgh: Creation Science Fellowship, 2003.

Notes

Chapter 1: Foundations of Baraminology

1. W. Greuter, J. McNeill, and F. R. Barrie, editors, *International Code of Botanical Nomenclature* (Port Jervis, N.Y.: Lubrecht & Cramer, Ltd.).

2. Wise, 1998.

3. Wise, 1998.

4. Biographical information is from the University of California Berkeley Museum of Paleontology web page: http://www.ucmp.berkeley.edu/history/evothought.html.

5. Material in this section is adapted from E. Mayr, *The Growth of Biological Thought* (Cambridge, Mass.: Belknap Press, 1982), pp. 149–152; and from S. Asma, *Following Form and Function* (Evanston, Ill.: Northwestern University Press, 1996).

6. Quoted in S. Asma, *Following Form and Function* (Evanston, Ill.: Northwestern University Press, 1996), p. 76.

7. Material in this section is adapted from J. Selden, "Aquinas, Luther, Melanchthon, and Biblical Apologetics," *Grace Theological Journal* 5, no. 2 (1984): 181–195, and from R. A. Muller, "Scholasticism, reformation, orthodoxy, and the persistence of Christian Aristotelianism," *Trinity Journal* 19, no. 1 (1998): 81–96.

8. Material in this section is adapted from E. Mayr, *The Growth of Biological Thought* (Cambridge, Mass.: Belknap Press, 1982), pp. 172–176.

9. J. Hunter, "Observations tending to show that the Wolf, Jackal, and Dog are of the same species", *Philosophical Transactions of the Royal Society of London* 77 (1787): 253–266.

10. W. Coleman, *Biology in the Nineteenth Century* (Cambridge: Cambridge University Press, 1977), p. 1.

11. S. Asma, *Following Form and Function* (Evanston, Ill.: Northwestern University Press, 1996).

12. C. Darwin, *The Origin of Species* (Reprint, New York: Gramercy Books, 1859), p. 455.

13. J. D. Hannah, "*Bibliotheca sacra* and Darwinism: An analysis of the nineteenth-century conflict between science and theology," *Grace Theological Journal* 4, no. 1 (1983): 37–58.

14. B. C. Nelson, *After its Kind* (Minneapolis: Augsburg Publishing, 1927), pp. 21ff.

15. W. Frair, "Frank Lewis Marsh–His Biography and His Baramins," in Helder, 2001, p. 9; and H. L. Armstrong, "Frank Lewis Marsh," *Creation Research Society Quarterly* 13 (1976): 3–4.

16. F. L. Marsh, 1941, p. 100.

17. F. L. Marsh, 1944, pp. 148–149.

18. F. L. Marsh, 1941, p. 92.

19. F. L. Marsh, 1976, p. 90.

20. F. L. Marsh, 1947, p. 177.

21. F. L. Marsh, 1944, p. 185.

22. J. C. Whitcomb, and H. M. Morris, *The Genesis Flood* (Phillipsburg, N.J.: Presbyterian & Reformed, 1961), pp. 66–67.

23. H. R. Siegler, "The magnificence of kinds as demonstrated by Canids," *Creation Research Society Quarterly* 11 (1974): 94–97.

24. A. J. Jones, "How many animals in the Ark?" *Creation Research Society Quarterly* 10 (1973): 102–108.

25. Scherer, 1993.

26. W. J. ReMine, "Discontinuity systematics: A new methodology of biosystematics relevant to the creation model," *Proceedings of the Second International Conference on Creationism,* editors R. E. Walsh and C. L. Brooks (Pittsburgh: Creation Science Fellowship, 1990), pp. 207–213.

27. Wise, 1990.

28. Frair, 2000.

29. See Robinson and Williams, 1999 and Helder, 2001.

Chapter 2: The Pattern of Life

1. E.g., see Marsh, 1947, p. 235.

2. Marsh, 1950.

3. Scherer, 1993.

4. S. Scherer, "Basic Types of Life," in Scherer, 1993, pp. 11–30.

5. The material on ReMine and Wise's baramin revision can be found in W. J. ReMine, "Discontinuity systematics: A new methodology of biosystematics relevant to the creation model," *Proceedings of the Second International Conference on Creationism,* editors R. E. Walsh and C. L. Brooks (Pittsburgh: Creation Science Fellowship, 1990), pp. 207–213; and in Wise, 1990.

6. The Baraminology Study Group, "The Refined Baramin Concept," *Occasional Papers of the Baraminology Study Group,* in preparation, 2003.

7. Wise, 1998.

8. Reynolds, 2002.

Chapter 3: The History of Baramins

1. NIDOTTE, #2012.

2. NIDOTTE, #6912.

3. G. J. Wenham, *Genesis 1–15* (Waco, Tex.: Word Books, 1987), pp. 20–21.

4. K. P. Wise, "Some thoughts on the Precambrian fossil record," *Creation Ex Nihilo Technical Journal* 6 (1992): 67–71.

5. NIDOTTE, #9490.

6. E.g., see the discussion in Westermann, 1994, p. 138.

7. NIDOTTE, #6416.

8. NIDOTTE, #2651.

9. NIDOTTE, #5883.

10. NIDOTTE, #989.

11. E.g., see H. M. Morris, *The Remarkable Record of Job* (Grand Rapids, Mich.: Baker Book House, 1988), pp. 115–117; and P. S. Taylor, *The Great Dinosaur Mystery and the Bible* (Colorado Springs: Chariot Victor Publishing, 1987), pp. 18–19.

12. NIDOTTE, #8253.

13. K. P. Wise, "Were there really no seasons? Tree rings and climate," *Creation Ex Nihilo Technical Journal* 6 (1992): 168–172.

14. This observation comes from Gary Phillips, Bryan College.

15. See discussion in Westermann, 1994, p. 237.

16. NIDOTTE, #5729.

17. Cited in G. J. Wenham, *Genesis 1–15* (Waco, Tex.: Word Books, 1987), p. 79.

18. See Westermann, 1994, p. 224.

19. NIDOTTE, #1998 and #7764.

20. NIDOTTE, #9300.

21. NIDOTTE, #7366.

22. See Wise and Croxton, 2003.

23. NIDOTTE, #5476.

24. Jones, 1973.

25. J. C. Whitcomb, and H. M. Morris, *The Genesis Flood* (Phillipsburg, N.J.: Presbyterian & Reformed, 1961), p. 69.

26. E.g., see O. Bar-Yosef, and M. E. Kislev, "Earliest domesticated barley in the Jordan valley," *National Geographic Research* 2, no. 2 (1986): 257.

27. S. A. Austin, A. A. Snelling, and K. P. Wise, "Canyon-length mass kill of orthocone nautiloids, Redwall Limestone (Mississippian), Grand Canyon, Arizona," *Geological Society of America Abstracts w/Programs* 31, no. 7 (1999): A421.

28. See S. A. Austin, "Depositional Environment of the Kentucky No. 12 Coal Bed (Middle Pennsylvanian) of Western Kentucky, with Special Reference to the Origin of Coal Lithotypes," Pennsylvania State University: Ph.D. dissertation 1979; and S. A. Austin, "Mount St. Helens and catastrophism," *Impact* (1996), 157.

29. M. M. M. Kuypers, P. Blokker, J. Erbacher, H. Kinkel, R. D. Pancost, S. Schouten, and J. S. Sinninghe Damsté, "Massive Expansion of Marine Archaea During a Mid-Cretaceous Oceanic Anoxic Event," *Science* 293 (2001): 92–95.

30. J. R. Baumgardner, "Runaway subduction as the driving mechanism for the Genesis Flood," *Proceedings of the Third International Conference on Creationism,* editor R. E. Walsh (Pittsburgh: Creation Science Fellowship, 1994), 63–79.

31. S. A. Austin, J. R. Baumgardner, D. R. Humphreys, A. A. Snelling, L. Vardiman, K. P. Wise, "Catastrophic Plate Tectonics: a global flood model of Earth history," *Proceedings of the Third International Conference on Creationism,* editor R. E. Walsh (Pittsburgh: Creation Science Fellowship, 1994), 609–621.

32. L. Vardiman, "A conceptual transition model of the atmospheric global circulation following the Genesis Flood," *Proceedings of the Third International Conference on Creationism,* editor R. E. Walsh, (Pittsburgh: Creation Science Fellowship, 1994), 569–579.

33. L. Vardiman, "Cooling of the Ocean After the Flood," *Impact* (1996): 277.

34. D. Jolly, and 32 others, "Biome reconstruction from pollen and plant macrofossil data for Africa and the Arabian peninsula at 0 and 6 ka," *Journal of Biogeography* 25 (1998): 1007–1027.

35. R. M. Schoch, and J. A. West, "Redating the Great Sphinx of Giza, Egypt," *Geological Society of America Abstracts With Programs* 23, no. 5 (1991): A253.

36. S. A. Austin, J. R. Baumgardner, D. R. Humphreys, A. A. Snelling, L. Vardiman, K. P. Wise, "Catastrophic Plate Tectonics: a global flood model of Earth history," *Proceedings of the Third International Conference on Creationism,* editor R. E. Walsh (Pittsburgh: Creation Science Fellowship, 1994), 609–621.

37. M. J. Oard, *An Ice Age Caused by the Genesis Flood* (El Cajon, Calif.: Institute for Creation Research, 1990); and L. Vardiman, *Ice Cores and the Age of the Earth* (El Cajon, Calif.: Institute for Creation Research, 1996).

38. See S. A. Austin, editor, *Grand Canyon: Monument to Catastrophe* (El Cajon, Calif.: Institute for Creation Research 1, 1994), pp. 92–107, and esp. note 40, p. 109.

39. See W. Ryan, and W. Pitman, *Noah's Flood: The New Scientific Discoveries about the Event that Changed History* (New York: Simon & Schuster, 1998). See also C. R. Froede, "Is the Black Sea Flood the Flood of Genesis?" *Creation Matters* 6, no. 1 (2001): 1–4.

40. S. A. Austin, "The Declining Power of Post-Flood Volcanoes," *Impact* (1998): 302.

41. D. D. Webb, "Faunal interchange between North and South America," *Acta Zoologica Fennica* 170 (1985): 177–178.

42. L. Vardiman, "A conceptual transition model of the atmospheric global circulation following the Genesis Flood," *Proceedings of the Third International Conference on Creationism,* editor R. E. Walsh (Pittsburgh: Creation Science Fellowship, 1994), 569–579.

43. Wise and Croxton, 2003.

44. Wood, Cavanaugh, and Wise, 2003; and K. P. Wise, "The camelidae fossil record," in Robinson and Williams, 1999, pp. 12–14.

45. See Wood, 2002b.

46. Wise and Croxton, 2003.

47. Wood, Cavanaugh, and Wise, 2003.

Chapter 4: Gathering and Interpreting Biblical Data

1. Much of the material in this chapter came from conversations with David Fouts, Bryan College, and some ideas were previously published in Wood, 2002a.

2. See P. J. Williams, "Biblical and Linguistic Studies on Camels in Baraminology," in Robinson and Williams, 1999, pp. 10–12.

3. See discussion in Wood, 2002a.

4. J. H. Hayes, and C. R. Holladay, *Biblical Exegesis: A Beginner's Handbook* (Atlanta: John Knox Press, 1987).

5. See NIDOTTE, #7366.

6. Wood, 2002a.

7. Information on translations in this section comes from P. W. Comfort, *The Complete Guide to Bible Versions* (Wheaton, Ill.: Living Books, 1991).

Chapter 5: Successive Approximation

1. Compare pp. 169, 181 with p. 160 of Marsh, 1947.

2. W. J. ReMine, "Discontinuity systematics: A new methodology of biosystematics relevant to the creation model," *Proceedings of the Second International Conference on Creationism,* editors R. E. Walsh, and C. L. Brooks (Pittsburgh: Creation Science Fellowship, 1990), 207–213.

3. Robinson and Cavanaugh, 1998b.

4. Scherer, 1993.

5. Wise, 1992.

6. Wood and Cavanaugh, 2001.

7. Wood and Cavanaugh, 2001; and Cavanaugh and Wood, 2002.

8. R. Junker, "Der Grundtyp der Weizenartigen (Poaceae, Tribus Triticeae)," in Scherer, 1993; and Wood, 2002a.

9. W. Frair, "Turtles: Now and then," *Proceedings of the Northcoast Bible-Science Conference* (Seaven Hills, Ohio: Bible Science Association 1984), 33–38; and W. Frair, "Original kinds and turtle phylogeny," *Creation Research Society Quarterly* 28 (1991): 21–24.

10. Wise, 1992.

11. Robinson, 1997.

Chapter 6: Identifying True Discontinuity

1. Wise, 1990.

2. Example taken from Wood and Cavanaugh, 2001.

3. Wise, 1990 and Wise, 1992.

4. Unless otherwise noted, these criteria are from Wise, 1990 and Wise, 1992.

5. W. J. ReMine, "Discontinuity systematics: A new methodology of biosystematics relevant to the creation model," *Proceedings of the Second International Conference on Creationism,* editors R. E. Walsh, and C. L. Brooks (Pittsburgh: Creation Science Fellowship, 1990), 207–213.

6. T. C. Wood, and D. P. Cavanaugh, "An Evaluation of Lineages and Trajectories as Baraminological Membership Criteria," *Occasional Papers of the Baraminology Study Group,* submitted 2002.

7. Robinson, 1997.

8. Wise, 1992.

9. W. Frair, "Some molecular approaches to taxonomy," *Creation Research Society Quarterly* 4(1) (1967): 18–22, 47. See also H. R. Wolfe, "Standardization of the precipitation technique and its application to studies of relationships in mammals, birds, and reptiles," *Biological*

Bulletin 76 (1939): 108–120; and E. Cohen, "Immunological studies of the serum proteins of some reptiles," *Biological Bulletin* 109 (1955): 394.

10. Robinson, 1997.

11. S. F. Gilbert, G. A. Loredo, A. Brukman, and A. C. Burke, "Morphogenesis of the turtle shell: the development of a novel structure in tetrapod evolution," *Evolution & Development* 3 (2001): 47–58.

12. Wise, 1992.

13. W. Frair, "Original kinds and turtle phylogeny," *Creation Research Society Quarterly* 28 (1991): 21–24.

Chapter 7: Hybridization

1. See J. F. Wendel, "Genome evolution in polyploids," *Plant Molecular Biology* 42 (2000): 225–249.

2. See Gray, 1954 and 1958.

3. J. Clausen, D. D. Keck, and W. M. Hiesey, "The concept of species based on experiment," *American Journal of Botany* 26 (1939): 103–106.

4. See W. S. Judd, C. S. Campbell, E. A. Kellogg, and P. F. Stevens, *Plant Systematics: A Phylogenetic Approach* (Sunderland, Mass.: Sinauer Associates, 1999), esp. p. 118ff.

5. Scherer, 1993.

6. E.g., see G. A. Marvin, and V. H. Hutchison, "Courtship behavior of the Cumberland Plateau woodland salamander, *Plethodon kentucki* (AmphibiaPlethodontidae), with a review of courtship in the genus *Plethodon*," *Ethology* 102 (1996): 285–303.

7. E.g., see E. Wiland, A. Wojda, M. Kamieniczna, M. Szczygiel, and M. Kurpisz, "Infertility status of male individuals with abnormal spermiogram evaluated by cytogenetic analysis and in vitro sperm penetration assay," *Medical Science Monitor* 8, no. 5 (2002): CR394–400.

8. Scherer, 1993.

9. J. Wells, "You can't get there from here: Discontinuity in development and evolution," in Helder, 2002, p. 24.

10. T. C. Wood, and 29 others, "HybriDatabase: A computer repository of organismal hybridization data," in Helder, 2002, p. 30.

11. I. W. Knobloch, *A Check List of Crosses in the Gramineae* (East Lansing, Mich.: published by the author, 1968).

12. Z. Kowalska, "A note on bear hybrids," *International Zoo Yearbook* 9 (1969): 89.

13. E.g., C. L. Hubbs, "Hybridization between fish species in nature," *Systematic Zoology* 4 (1955): 1–20.

14. Thanks to Stephanie Mace for teaching us this trick.

15. J. Fehrer, "Interspecies-kreuzungen bei carduelíden finken und prachtfinken," in Scherer, 1993, pp. 197–215.

16. Wood, 2002.

17. Robinson, 1997.

18. Robinson and Cavanaugh, 1998b.

19. S. Scherer, "Der grundtyp der entenartigen (Anseriformes, Anatidae): Biologische und paläontologische streiflichter," in Scherer, 1993, pp. 131–158.

20. Wood and Cavanaugh, 2001.

21. E. More, "The created kind: Noah's doves, ravens, and their descendants," *Proceedings of the Second International Conference on Creationism,* editor R. E. Walsh (Pittsburgh: Creation Science Fellowship, 1998), 407–419.

Chapter 8: Statistical Baraminology

1. Robinson and Cavanaugh, 1998a.

2. R. V. Sternberg, and D. P. Cavanaugh, "Analysis of morphological groupings using ANOPA, a pattern recognition and multivariate statistical method: A case study involving centrarchid fishes," manuscript submitted.

3. E.g., see Wood, 2002.

4. Baraminic distance methodology was originally published in Robinson and Cavanaugh, 1998a and 1998b.

5. Wood, 2002.

6. Robinson and Cavanaugh, 1998a.

7. Robinson and Cavanaugh, 1998b.

8. Wood, 2002.

9. Wood, Cavanaugh, and Wise, 2003.

10. S. Hartwig-Scherer, "Hybridisierung und Artbildung bei den Meerkatzenartigen (Primates, Cercopithecoidea)," in Scherer, 1993, pp. 245–257.

11. R. Junker, "Der Grundtyp der Weizenartigen (Poaceae, Tribus Triticeae)," in Scherer, 1993, pp. 75–93.

12. G. F. Howe, and J. R. Meyer, "The growth rate of *Muhlenbergia torreyi* (Ring Muhly grass) colonies in central Arizona," *Creation Research Society Quarterly* 28 (2001): 159–161.

13. R. V. Sternberg, and D. P. Cavanaugh, "Analysis of morphological groupings using ANOPA, a pattern recognition and multivariate statistical method: A case study involving centrarchid fishes," manuscript submitted.

14. Cavanaugh and Wood, 2002.

15. K. Bremer, *Asteraceae Cladistics & Classification* (Portland, Oreg.: Timber Press, 1994), p. 43.

16. P. O. Karis, "*Heliantheae* sensu lato (*Asteraceae*), clades and classification," *Plant Systematics and Evolution* 188 (1993): 139–195.

17. T. C. Wood, and D. P. Cavanaugh, "An Evaluation of Lineages and Trajectories as Baraminological Membership Criteria," *Occasional Papers of the Baraminology Study Group,* submitted 2002.

18. Wood, Cavanaugh, and Wise, 2003.

19. T. C. Wood, and D. P. Cavanaugh, "An Evaluation of Lineages and Trajectories as Baraminological Membership Criteria," *Occasional Papers of the Baraminology Study Group,* submitted 2002.

20. Cavanaugh and Wood, 2002.

Chapter 9: Baraminology and Design

1. See W. A. Dembski, ed., *Mere Creation* (Downers Grove, Ill.: InterVarsity Press, 1998).

2. C. B. Thaxton, W. L. Bradley, and R. L. Olsen, *The Mystery of Life's Origin* (Dallas: Lewis and Stanley, 1984).

3. S. C. Meyer, "The explanatory power of design," in *Mere Creation,* editor W. A. Dembski (Downers Grove, Ill.: InterVarsity Press, 1998), 113–147.

4. W. Paley, *Natural Theology,* 12th ed. (London: J. Faulder, 1802), p. 456ff.

5. W. Paley, *Natural Theology,* 12th ed. (London: J. Faulder, 1802), p. 468ff, esp. p. 471.

6. Marsh, 1947, p. 225.

7. H. W. Clark, "Paleoecology and the Flood," *Creation Research Society Quarterly* 8 (1971): 19–23.

Chapter 10: Biological Imperfection

1. S. J. Gould, *The Panda's Thumb* (New York: Norton, 1980), pp. 19–26.

2. S. J. Gould, *The Panda's Thumb* (New York: Norton, 1980), p. 20.

3. See references in T. C. Wood, "Genome decay in the mycoplasmas," *Impact* (2001): 340.

4. T. E. Dowling, D. P. Martasian, and W. R. Jeffery, "Evidence for Multiple Genetic Forms with Similar Eyeless Phenotypes in the Blind Cavefish, *Astyanax mexicanus,*" in *Molecular Biology and Evolution* 19 (2002):446–455.

5. International Human Genome Sequencing Consortium, "Initial sequencing and analysis of the human genome," *Nature* 409 (2001): 860–921.

6. E.g., see T. H. Goldsmith, "Optimization, constraint, and history in the evolution of eyes," *The Quarterly Review of Biology* 65 (1990): 281–322; and G. Ayoub, "On the design of the vertebrate retina," *Origins & Design* 17, no. 1 (1996): 19–22.

7. R. Dawkins, *The Blind Watchmaker,* new edition (New York: Norton, 1996), p. 93.

8. M. Mwanatambwe, N. Yamada, S. Arai, M. Shimizu-Suganuma, K. Schichinohe, and G. Asano, "Ebola hemorrhagic fever (EHF): Mechanism of transmission and pathogenicity," *Journal of Nippon Medical School* 68, no. 5 (2001): 370–375.

9. See M. Balter, "On the trail of Ebola and Marburg Viruses," *Science* 290 (2000): 923–925.

10. See Y. Suzuki, and T. Gojobori, "The origin and evolution of Ebola and Marburg viruses," *Molecular Biology and Evolution* 14 (1997): 800–806.

11. M. Mwanatambwe, N. Yamada, S. Arai, M. Shimizu-Suganuma, K. Schichinohe, and G. Asano, "Ebola hemorrhagic fever (EHF): Mechanism of transmission and pathogenicity," *Journal of Nippon Medical School* 68, no. 5 (2001): 370–375; and M. Balter, "On the trail of Ebola and Marburg Viruses," *Science* 290 (2000): 923–925.

12. M. E. Miranda, T. G. Ksiazek, T. J. Retuya, A. S. Khan, A. Sanchez, C. F. Fulhorst, P. E. Rollin, A. B. Calaor, D. L. Manalo, M. C. Roces, M. M. Dayrit, and C. J. Peters, "Epidemiology of Ebola (subtype Reston) virus in the Philippines, 1996," *Journal of Infectious Diseases* 179, Suppl. 1 (1999): S115–119.

13. E. J. Rayfield, D. B. Norman, C. C. Horner, J. R. Horner, P. M. Smith, J. J. Thomason, and P. Upchurch, "Cranial design and function in a large theropod dinosaur," *Nature* 409 (2001): 1033–1037.

14. K. Chin, T. T. Tokaryk, G. M. Erickson, and L. C. Calk, "A king-sized theropod coprolite," *Nature* 393 (1998): 680–682.

15. See S. Perkins, "Turn your head and roar," *Science News* 160, no. 24 (2001): 376.

16. See S. Perkins, "Turn your head and roar," *Science News* 160, no. 24 (2001): 376.

17. This section is based on T. C. Wood, "The terror of anthrax in a degrading creation," *Impact* (2002): 345.

Chapter 11: Baraminology and Diversification

1. *Diversification* as a technical term was first recommended and used by Kurt Wise (e.g., see K. P. Wise, "North American Paleontology Convention96," *Creation Ex Nihilo Technical Journal* 10 (1996): 315–321.

2. Wise, 1990.

3. See Wood, 2002b.

4. See T. C. Wood, P. J. Williams, K. P. Wise, and D. A. Robinson, "Summaries on camel baraminology," in Robinson and Williams, 1999, pp. 9–20.

5. Wood, Cavanaugh, and Wise, 2003.

6. See Robinson and Cavanaugh, 1998b; and Crompton, N. E. A., "A review of selected features of the family Canidae with reference to its fundamental taxonomic status," in Scherer, 1993, pp. 217–224.

7. F. Zimbelmann, "Grundtypen bei Greifvögeln (Falconiformes)," in Scherer, 1993, pp. 185–195.

8. E.g., see J. E. Sulston, E. Schierenberg, J. G. White, and J. N. Thomson, "The embryonic cell lineage of the nematode *Caenorhabditis elegans*," *Developmental Biology* 100, no. 1 (1983): 64–119.

9. These characteristics are adapted from Wood, 2002b.

10. The term *heterozygous fractionation* was introduced in Wood, 2002b. See also C. Wieland, "Variation, information and the created kind," *Creation Ex Nihilo Technical Journal* 5 (1991): 42–47.

11. See Wood, 2002b; and Wood, 2003.

12. T. D. Read, and 24 others, "Genome sequences of *Chlamydia trachomatis* MoPn and *Chlamydia pneumoniae* AR39," *Nucleic Acids Research* 28 (2000): 1397–1406.

13. Wood, 2002a.

14. E.g., see M. D. Gale, and K. M. Devos, "Comparative genetics in the grasses," *Proceedings of the National Academy of Sciences USA* 95 (1998): 1971–1974.

15. R. Kalendar, J. Tanskanen, S. Immonen, E. Nevo, and A. H. Schulman, "Genome evolution of wild barley (*Hordeum spontaneum*) by *BARE*-1 retrotransposon dynamics in response to sharp microclimatic divergence," *Proceedings of the National Academy of Sciences USA* 97 (2000): 6603–6607.

16. Material from this section comes from F. Bushman, *Lateral DNA Transfer* (Cold Spring Harbor, N.Y.: CSHL Press, 2002), particularly chapter 3.

Chapter 12: Baraminology and Biogeography

1. See H. W. Clark, "Paleoecology and the Flood," *Creation Research Society Quarterly* 8 (1971): 19–23.

2. See H. Coffin, *Origins by Design* (Hagerstown, Md: Review and Herald Publishing, 1983), pp. 71ff; L. Brand, *Faith, Reason, & Earth History* (Berrien Springs, Mich.: Andrews University Press, 1997), p. 281; A. A. Roth, *Origins* (Hagerstown, Md: Review and Herald Publishing, 1998), pp. 170ff.

3. This model comes from conversations with Kurt Wise, Bryan College.

4. K. P. Wise, "The hydrothermal biome: A pre-Flood environment," *Proceedings of the Fifth International Conference on Creationism,* editor R. E. Walsh (Pittsburgh: Creation Science Fellowship, 2003).

5. S. A. Austin, and K. P. Wise, "The pre-Flood/Flood boundary: as defined in Grand Canyon, Arizona and eastern Mojave desert, California," *Proceedings of the Third International Conference on Creationism,* editor R. E. Walsh (Pittsburgh: Creation Science Fellowship, 1994), 37–47.

6. K. P. Wise, "The pre-Flood floating forest: A study in paleontological pattern recognition," *Proceedings of the Fifth International Conference on Creationism,* editor R. E. Walsh (Pittsburgh: Creation Science Fellowship, 2003).

7. J. R. Baumgardner, "Runaway subduction as the driving mechanism for the Genesis Flood," *Proceedings of the Third International Conference on Creationism,* editor R. E. Walsh (Pittsburgh: Creation Science Fellowship, 1994), 63–75.

8. Sections 12.3 and 12.4 are adapted from Wise and Croxton, 2003.

9. See S. A. Austin, *Mt. St. Helens: Explosive Evidence for Catastrophe* (El Cajon, Calif.: Institute for Creation Research, 1995).

10. H. G. Coffin, "The Yellowstone petrified 'forests,'" *Origins* 24 (1997): 5–44.

11. See B. J. MacFadden, *Fossil Horses* (New York: Cambridge University Press, 1992).

12. M. van Tuinen, C. G. Sibley, and S. B. Hedges, "Phylogeny and biogeography of ratite birds inferred from DNA sequences of the mitochondrial ribosomal genes," *Molecular Biology and Evolution* 15 (1998): 370–376.

13. See G. E. Schatz, "Malagasy/Indo-Australo-Malesian phytogeographic connections," *Biogeography of Madagascar,* editor W. R. Lourenço (Paris: Editions ORSTOM, 1996).

14. For a discussion of mammalian biogeography, see C. B. Cox, "Plate tectonics, seaways and climate in the historical biogeography of mammals," *Memórias do Instituto Oswaldo Cruz* 95 (2000): 509–516.

15. S. D. Webb, "Faunal interchange between North and South America," *Acta Zoologica Fennica* 170 (1985): 177–178.

16. See D. M. Christie, R. A. Duncan, A. R. McBirney, M. A. Richards, W. M. White, K. S. Harpp, and C. G. Fox, "Drowned islands downstream from the Galápagos hotspot imply extended speciation times," *Nature* 355 (1992): 246–248.

17. P. R. Grant, "Hybridization of Darwin's finches on Isla Daphne Major, Galápagos," *Philosophical Transactions of the Royal Society of London,* Series B 340 (1993): 127–139.

18. K. Rassmann, F. Trillmich, and D. Tautz, "Hybridization between the Galápagos land and marine iguana (*Conolophus subcristatus and Amblyrhynchus cristatus*) on Plaza Sur," *Journal of Zoology, London* 242 (1997): 729–739.

19. A. Sato, H. Tichy, C. O'hUigin, P. R. Grant, B. R. Grant, and J. Klein, "On the origin of Darwin's finches," *Molecular Biology and Evolution* 18 (2001): 299–311.

20. A. Caccone, J. P. Gibbs, V. Ketmaier, E. Suatoni, J. R. Powell, "Origin and evolutionary relationships of giant Galápagos tortoises," *Proceedings of the National Academy of Sciences USA* 96 (1999): 13223–13228.

21. A. Caccone, J. P. Gibbs, V. Ketmaier, E. Suatoni, J. R. Powell, "Origin and evolutionary relationships of giant Galápagos tortoises," *Proceedings of the National Academy of Sciences USA* 96 (1999): 13223–13228.

22. S. Fridriksson, and B. Magnússon, "Development of the ecosystem on Surtsey with references to Anak Krakatau," *GeoJournal* 28, no. 2 (1992): 287–291.

Glossary

adaptation. As a verb, adaptation is the process by which the anatomy of an organism becomes fit to a particular environmental condition. As a noun, an adaptation is any anatomical feature of an organism that is particularly suited to performing a function in the organism's environment.

additive evidence. Data used to demonstrate continuity (*q.v.*) between organisms. See also SUBTRACTIVE EVIDENCE.

analogy. According to Owen, analogy is any similarity of function between two anatomical structures. In modern evolutionary biology, analogy is any similarity that is independently derived (i.e., not inherited from a common ancestor).

Analysis of Patterns (ANOPA). A statistical method devised by Cavanaugh for analyzing multidimensional data in three dimensions.

apobaramin. According to the refined baramin concept, an apobaramin is a group of organisms discontinuous (*q.v.*) with all others.

apomixes. The elimination of all chromosomes from one parent during embryonic development.

baramin. According to the refined baramin concept, a baramin is the actualization of the potentiality region (*q.v.*) at any point or period of history.

baraminic distance. The fraction of characters (*q.v.*) that two organisms have in common. Also, a modern statistical method of evaluating baraminic membership.

baraminic signal. A modified chi-square statistic to measure the information content of an organismal dataset for use in a baraminic distance (*q.v.*) analysis.

baraminology. A creationist biosystematic (*q.v.*) method proposed by Wise in 1990. Baraminology combines the terminology and methodology of discontinuity systematics (*q.v.*) with explicitly creationist assumptions and criteria.

basic type biology. A biosystematic (*q.v.*) method proposed by Scherer in 1990. Basic type biology uses only hybridization to determine membership in basic types (Marshian baramins).

binomial nomenclature. The system of naming species introduced by Linnaeus. Each species is described both by the genus and species name, typically Latin or Latinized words.

biogeography. The branch of biology that deals with the spatial distribution of organisms.

biological character space. A theoretical, multidimensional graph of all possible attributes of all possible organisms.

biological imperfection. An attribute of an organism that appears to contradict God's original intention for creation.

biological trajectory. A set of organisms that describe a linear path through biological character space (*q.v.*).

biosystematics. The science of discovery and organization of biological diversity.

center of survival. A geographical point from which a group of organisms survived the Flood and radiated during the post-Flood period. The center of survival for land organisms would be Ararat.

chain of being. Aristotle's view of the classification of organisms, in which they are arranged in a line from "lower" organisms (flies, worms) to "higher" organisms (human males).

character. A particular attribute possessed by a group of organisms. Characters may have unlimited values (continuous characters), a limited set of values (multistate discrete characters) or just two values (binary discrete characters).

character diversity. A measurement of variation in an organismal dataset used in baraminic distance (*q.v.*) analysis.

character relevance. In a baraminic distance (*q.v.*) analysis, the percentage of organisms for which a character state (*q.v.*) is known.

character state. The actual value of a character (*q.v.*) for a particular organism.

comparium. According to Clausen, Keck, and Hiesey, a comparium is a group of organisms that are capable of hybridizing. The comparium roughly corresponds with Marsh's baramin (*q.v.*) or Scherer's basic type (*q.v.*).

consilience of induction. A theory capable of explaining many diverse types of data that might otherwise be considered unrelated.

continuity. According to the refined baramin concept (*q.v.*), continuity is significant, holistic similarity.

creation. The model of origins that assumes that God created the universe in a mature state, six thousand years ago, cursed it because of human sin, sent a global Flood to punish human wickedness, and confused human languages at the historical tower of Babel.

discontinuity. According to the refined baramin concept, discontinuity is significant, holistic difference between two organisms or groups of organisms.

discontinuity systematics. A biosystematics method proposed by ReMine in 1990. Discontinuity systematics emphasizes the approximation of the holobaramin (*q.v.*) by the accumulation of additive and subtractive evidence (*q.v.*).

diversification. A rapid process by which a baramin becomes stable species.

endemism. The occurrence of multiple species in a particular region but nowhere else.

essentialism. A view of biology, endorsed by Aristotle, in which organisms may be classified as species by the possession of certain immutable and defining attributes.

evolution. The model of origins that assumes that all organisms on earth descended from a common ancestor by natural processes of variation and selection.

exegesis. The process of interpretation, from a Greek word meaning "to draw out."

fixity. A corollary of essentialism (*q.v.*) which posits that species cannot change.

functionalism. A popular model of biology that seeks to understand biological phenomena in terms of the processes in which they participate. See also STRUCTURALISM.

genomic modularity. A mechanism of diversification (*q.v.*) proposed by Wood in 2003. Genomic modularity is believed to work by remodeling the genome, possibly in response to stress.

heterosis. The condition of an interspecific hybrid that is more fit or hearty than either parent. Also called HYBRID VIGOR.

heterozygous fractionation. A mechanism of diversification (*q.v.*) popular among modern creationists. Heterozygous fractionation assumes that God created all genetic information for a baramin (*q.v.*) in the form of alleles. Diversification proceeds by the assortment of alleles to particular lineages.

holobaramin. According to the refined baramin concept, a holobaramin is a group of organisms within which every member shares continuity with at least one other member, but no member is continuous with any organisms outside the holobaramin.

homology. Originally defined by Owen as "the same organ in different animals," homology has come to mean structures or genes that share a common ancestor.

hybrid swarm. A group of (often sterile) first-generation, interspecific hybrids found in a limited geographic region.

hybrid vigor. See HETEROSIS.

hybridization. The result of mating between different organisms, often referring to the mating of organisms of different species. The offspring of hybridization are called hybrids.

hybridization network. A graphical tool used to illustrate hybridization potential between groups of organisms.

hybridogram. A graphical tool used to illustrate known hybridization between particular species.

introgression. Alleles from one species that occur in the gene pool of another as a result of interbreeding and the production of fertile interspecific hybrids.

macroevolution. An ill-defined term that describes evolution above the level of species. Because of the confusion regarding the precise definition of this term, it should be avoided.

mediated design. A mechanism by which an intention of God is accomplished by another agent or a group or series of agents.

microevolution. Properly, all variation that occurs within a species.

monobaramin. According to the refined baramin concept, a monobaramin is a group of organisms within which all members share continuity (*q.v.*) with at least one other member.

natural evil. A term used to designate all events and phenomena that cause death and suffering but are not the result of human intention.

natural theology. 1. A school of thought that sought to construct theology through analysis of creation and application of reason. 2. The title of a book published by William Paley that is regarded as the definitive statement of natural theology.

nomenclature. In biosystematics, the formal procedure by which a species or larger group is given a name.

parsimony. A principle of cladistics and evolution that posits that the evolutionary tree with the least changes and inconsistencies is most likely correct.

potentiality region. A region of biological character space (*q.v.*) within which any combination of character states (*q.v.*) could produce a viable organism, but outside of which, no combination of character states (*q.v.*) could produce a viable organism.

reductionism. A common paradigm in modern biology that seeks to understand biological phenomena by examining their constituent parts. Implicitly, reductionists believe that biology at any level can be explained by the accumulated action of lower levels of organization.

refined baramin concept. A definition of baramin (*q.v.*) and associated terms proposed and advanced by the Baraminology Study Group in 2002. The refined baramin concept seeks to assimilate all positive and useful features of previous baramin concepts.

Scala Naturae. 1. The title of Aristotle's book explaining the chain of being (*q.v.*). 2. Another term for the chain of being.

scholasticism. A school of theology which believed that knowledge of God could be acquired by the application of human reason alone.

speciation. The process by which one species gives rise to another. Because speciation has become so closely associated with neodarwinism, baraminologists recommend using the term diversification (*q.v.*).

species. A lineage of a holobaramin (*q.v.*) that is stable in morphology and interfertile in practice.

species concept. An attempt to rigorously define species (*q.v.*). Many species concepts have been proposed over the years.

structuralism. A school of anatomy that seeks to explain the form of an organism without recourse to function. See also FUNCTIONALISM.

subtractive evidence. Data used to demonstrate discontinuity (*q.v.*) between organisms. See also ADDITIVE EVIDENCE.

successive approximation. The process of estimating the membership of the holobaramin (*q.v.*) through gathering of additive and subtractive evidence. These estimates are continually refined as more evidence is gathered.

synapomorphy. In cladistics, a character state shared by a group of organisms that was first derived in their ancestor.

taxon. A group of organisms classified in taxonomy.

taxonomy. The classification of organismal diversity.

theistic evolution. A model of origins that assumes that God used the process of evolution (*q.v.*) to create living organisms.

transformism. A school of biology that believes that a species is capable of becoming a different species.

Trinity. The doctrine formulated by the Council of Nicea that attempts to describe the nature of God as three persons in one God.

typology. A model of structuralism advocated by Sir Richard Owen in which the organization of biological form represents divine plan.

universal ancestry. A corollary of modern evolution (*q.v.*) that posits that all organisms on earth descended from a single ancestor.

vicariance. A model of biogeography that suggests that much of organismal dispersion occurred passively as a result of continental drift.

Index

Credits

All figures are used by permission of the copyright owner or are public domain. Figures 2.2 and 11.1 include images from Corel Draw 9.0, which are protected by the copyright laws of the U.S., Canada, and elsewhere. Used under license.

1.1: T. Wood
1.3: Perry-Castañeda Library, University of Texas at Austin
1.4: Perry-Castañeda Library, University of Texas at Austin
1.5: Adventist Heritage Center, Andrews University, used by permission
2.1: T. Wood and K. Wise
3.3: Bryan College
3.4: NASA
5.5: USGS and Caribbean Conservation Corporation
7.2: Copyright 1993 Pascal-Verlag, used by permission
7.3: Copyright 2002 Answers in Genesis, used by permission
7.4: Copyright 1997 Creation Research Society, used by permission
7.5: Copyright 1998 Creation Research Society, used by permission
7.6: Copyright 1993 Pascal-Verlag, used by permission
7.7: Copyright 2001 Geoscience Research Institute, used by permission
7.8: Copyright 1998 Creation Research Society, used by permission
8.4: Copyright 2002 Baraminology Study Group, used by permission
8.5: Copyright 2002 Baraminology Study Group, used by permission
9.5: Stephen Low Productions
10.3: T. Allison
11.3: T. Wood
11.4: T. Allison
12.2: USGS
12.3: USGS
12.4: D. Murray
12.5: NASA
12.6: T. Wood and S. Mace